AF595119

Advances in Plasma Diagnostics and Applications

Advances in Plasma Diagnostics and Applications

Editors

Zhitong Chen
Pankaj Attri
Qiu Wang

MDPI • Basel • Beijing • Wuhan • Barcelona • Belgrade • Manchester • Tokyo • Cluj • Tianjin

Editors

Zhitong Chen
Shenzhen Institute of
Advanced Technology
China

Pankaj Attri
Kyushu University
Japan

Qiu Wang
Institute of Mechanics
China

Editorial Office
MDPI
St. Alban-Anlage 66
4052 Basel, Switzerland

This is a reprint of articles from the Special Issue published online in the open access journal *Processes* (ISSN 2227-9717) (available at: https://www.mdpi.com/journal/processes/special_issues/Plasma_Diagnostics).

For citation purposes, cite each article independently as indicated on the article page online and as indicated below:

LastName, A.A.; LastName, B.B.; LastName, C.C. Article Title. *Journal Name* **Year**, *Volume Number*, Page Range.

ISBN 978-3-0365-4319-2 (Hbk)
ISBN 978-3-0365-4320-8 (PDF)

Contents

About the Editors

Zhitong Chen

Zhitong Chen is a professor at the Advanced Therapy Center at National Innovation Center for Advanced Medical Devices. He founded an iPlasma Research Lab and served as director of the Advanced Therapy Center. His iPlasma Lab-related research focuses on plasma science and plasma-related applications: biomedicine, medical device, and propulsion. He completed his postdoctoral training in the Plasma and Space Propulsion Lab of Dr. Richard E. Wirz at the University of California, Los Angeles. He received his Ph.D. in Mechanical and Aerospace Engineering from George Washington University in 2017, where he worked under the supervision of Dr. Michael Keidar. He received his M.Sc. and B.Sc. in Engineering Mechanics at the Institute of Mechanics, Chinese Academy of Sciences, and Mechanical Engineering at Northeast Agriculture University, respectively. He has (co)authored over 40 papers (including PNAS and Sci. Adv.), one book, over 20 patents, and over 30 conference contributions.

Pankaj Attri

Pankaj Attri joined Kyushu Univ. in 2019, as an Associate Professor. He received his Master's and and Ph.D. degree in Chemistry from Univ. of Delhi, India. He worked as Assistant Professor in Kwangwoon Univ., Korea from 2013–2017. He was awarded by Japan Society for the Promotion of Science invitation fellowship at Kyushu Univ. in 2016, and received an Marie Skłodowska-Curie actions individual fellowship at Antwerp Univ., Belgium in 2017. Dr Attri has over 120 publications, 10 book chapters and 5 patents. Dr Attri is working as an Editorial Board Member of Scientific Reports and Magnetochemistry Journals and also working as a guest editor in various journals.

Qiu Wang

Qiu Wang is an Associate Professor in the Institute of Mechanics, Chinese Academy of Sciences (CAS). He received his Bachelor's degree in School of Energy and Power Engineering in Beihang University in 2008, and PhD in Fluid Mechanics from the Institute of Mechanics, Chinese Academy of Sciences in 2013. Sponsored by the China Scholarship Council, he completed a one-year stay as a visiting scholar in RWTH Aachen University in Germany. His current research interests include nonequilibrium flows, hypersonic aerothermal dynamics, hypersonic magnetohydrodynamic flows, and oblique detonation propulsion. He has undertaken several projects as a project leader, including the National Natural Science Foundation of China (Grant No. 12072352) and the Youth Innovation Promotion Association CAS (Grant No. 2021020). He has published more than 30 papers and been granted 9 Chinese patents. He is a youth editorial board member of Aerodynamic Research & Experiment. He also received the Science & Technology Progress Award from the Chinese Society of Theoretical and Applied Mechanics in 2021.

Preface to "Advances in Plasma Diagnostics and Applications"

Plasma is the fourth state of matter, and is contrasted with the other states: solid, liquid, and gas. Plasma can exist in various forms due to discharge modes that are created in different ways, resulting in a wide range of applications. Plasma applications are interdisciplinary research fields that combine physics, chemistry, biology, and medicine. When plasma comes into contact with materials, it generates intense UV radiation, reactive species, electrons, and charged particles, all of which are effective agents against many matters, and their processes are incredibly complex. However, the plasma parameter space is extensive and ranges from the density of radicals to the velocity distribution of charged particles, and even rovibrationally excited states or the population of excited electronics. The aim of this Special Issue is to collect original research and review articles on the most recent plasma applications and diagnose their processes, to elucidate the characteristics of plasma and the mechanisms of plasma-induced processes. Potential topics include, but are not limited to:

Experimental measurement methods and diagnostic tools for the detection of radicals in the gas and liquid phase;

Modification in materials;

Materials processing (deposition, etching, washing, etc.);

Industrial applications including water purification;

Biomedical/Agricultural/Food processing applications.

Zhitong Chen, Pankaj Attri, and Qiu Wang
Editors

 MDPI

Editorial

Special Issue on "Advances in Plasma Diagnostics and Applications"

Zhitong Chen [1,2,*], Pankaj Attri [3,4,*] and Qiu Wang [5,*]

1 Center for Advanced Therapy, National Innovation Center for Advanced Medical Devices, Shenzhen 518000, China
2 Institute of Biomedical and Health Engineering, Shenzhen Institute of Advanced Technology, Chinese Academy of Sciences, Shenzhen 518055, China
3 Center of Plasma Nano-interface Engineering, Kyushu University, Fukuoka 819-0395, Japan
4 Graduate School of Information Science and Electrical Engineering, Kyushu University, Fukuoka 819-0395, Japan
5 State Key Laboratory of High Temperature Gas Dynamics, Institute of Mechanics, Chinese Academy of Sciences, Beijing 100190, China
* Correspondence: zt.chen@nmed.org.cn (Z.C.); attri.pankaj.486@m.kyushu-u.ac.jp (P.A.); wangqiu@imech.ac.cn (Q.W.)

Citation: Chen, Z.; Attri, P.; Wang, Q. Special Issue on "Advances in Plasma Diagnostics and Applications". *Processes* **2022**, *10*, 654. https://doi.org/10.3390/pr10040654

Received: 14 February 2022
Accepted: 25 March 2022
Published: 28 March 2022

Publisher's Note: MDPI stays neutral with regard to jurisdictional claims in published maps and institutional affiliations.

Plasma is the fourth state of matter, contrasted with the other states: solid, liquid, and gas. It is the most energetic and abundant state of matter, with as much as 99.9% of the Universe's matter composed of plasma. Plasmas can be generated via the combination of energy-induced fragmentation, ionization, and excitation of atoms and molecules, resulting in a vast variety of atomic and molecular species, which can be electrically charged, energetically excited, highly reactive, or any combination of these states. The basic plasma parameters are very broad and cover the density of radicals and charged particles; i.e., electrons and positive and negative ions are subject to their velocity distribution, even in rovibrationally excited electronic states.

Since World War II, many different methods for generating plasmas have been developed. Starting with microwave-driven discharges, a spin-off of radar, in the 1950s, the generation of radio-frequency came into focus in the 1960s, typically in a parallel-plate reactor of a capacitively coupled plasma (CCP), and in the 1980s, the separate control of plasma density and the energy of the ions was the main issue, in terms of discharge, which can be subsidized as electron cyclotron resonance (ECR) and inductively coupled plasmas (ICP).

Subsequently, an array of methods for plasma diagnostics was developed. With the electric probe of Langmuir, only measurements of plasma density and electron temperature in plasmas through inert gases (argon, nitrogen, oxygen) were feasible. But today, investigations into all different types of plasmas, irrespective of being inert or aggressive, are performed. Among them are optical measurements, in particular optical emission spectroscopy (OES) and optical absorption spectroscopy, in the infrared, scattering techniques (such as Rayleigh-, Raman- and Thomson scattering), mass spectrometry, electron paramagnetic resonance spectroscopy, gas chromatography, and various others. Also, the plasma itself became a research topic by self-excited electron resonance spectroscopy (SEERS) and measuring the complex impedance by recording the V(I) characteristics, just to name a few.

Although various mature diagnostic technologies for plasma discharges have been developed, there are still many challenges. A number of things must be done for full knowledge and understanding of plasmas, in order to guide without any break. The measurement precision is not only affected by the diagnostic equipment/techniques, but also the plasma discharge itself. In many applications, direct measurements of the parameters of interest are still not possible. In addition, the plasma environments in application processes are unusually complex, and their reactions are still not fully understood. Plasmas bring together many research fields, including statistical physics, atomic/molecular physics, thermodynamics, electrodynamics, fluid dynamics, heat transfer, electrical engineering, material, chemical engineering, surface science, physics, chemistry, and more recently, aerospace, agriculture, foods, environment, energy, catalysis, biology, and medicine.

This Special Issue on "Advances in Plasma Diagnostics and applications" of *Processes*, collects original research and review articles on the most recent plasma applications and the diagnosis of their processes, to elucidate the characteristics of plasma and mechanisms of plasma-induced processes. Wang et al. introduced the cowl-induced expansion wave, based on the model with an upper-side expansion wall, and analyzed the oblique detonation wave (ODW) dynamics, employing the reactive Euler equations, with a two-step induction–reaction kinetic model [1]. The numerical results show that the upstream movement of Mach stem can be re-stabilized for the unstable structure, which suggests a feasible adjustment method and the corresponding transient phenomena deserve more attention in future work. Xie and coworkers experimentally investigated the control of ramp shock wave in Mach 3 supersonic flow, using a two-electrode Spark Jet (SPJ) actuator, through schlieren images and dynamic pressure measurement results [2]. The ramp pressure is reduced by a maximum of 79%, compared to the pressure in the base flow field. The control effect on the ramp shock, in the case of medium ramp distance, is better when the SPJ exit is located outside the separation zone. Chen et al. characterized the gliding arc plasma, through periodic discharge, current, voltage, and power waveforms, as well as plasma topology related to the air flow rate [3]. The extinction performance of the flame was influenced, deeply, by the static flame instability. Due to the gliding arc plasma adopted, the lean blowout limit of the swirl flame in pulsating flow mode was significantly reduced and was found to be better than the limit of the stable flame.

Xu et al. deposited an Fe-based amorphous alloy, reinforced WC-Co-based coating, on 42CrMo steel, using plasma spray welding [4]. The coatings have a full metallurgical bond in the coating/substrate interface and the powder composition plays an important role in the micro-structures and properties of the coating. The interface, between the spray welding layer and matrix, has a large amount of WC, with a gradient distribution from internal to external in the cross section. The spray welding layer enhances the wear resistance and hardness of the 42CrMo steel.

Hwang and coworkers successfully synthesized the size-controlled carbon nanoparticles (CNPs), using the Ar +CH_4 MHD-PECVD (magnetohydrodynamics-plasma-enhanced chemical vapor deposition) method continuously [5]. The size of CNPs was proportional to the gas residence time in the discharges maintained in the electrode's holes, and the control range of the mean size was between 25 and 220 nm. They confirmed the deposition of carbon-related radicals, as the dominant process for nanoparticle growth processes in plasmas. The radical deposition developed the nucleated nanoparticles during the discharges' transport of CNPs, and the time of flight in discharges controlled the size of the nanoparticles.

Chen, Obenchain, and Wirz developed a single-electrode tiny plasma device to overcome the drawbacks of conventional plasma jets and investigated its physics and interactions with five subjects (DI water, metal, cardboard, belly, and muscle) [6]. For non-conductive subjects, reactive oxygen and nitrogen species (RONS) intensity shows very little change, with distance decreasing from 15 mm to 10 mm, while RONS intensity increases for conductive subjects, with distance decreasing, especially for the muscle. The center temperature of jet–subjects interaction still remains in the comfortable temperature range for human beings, after 2 min interaction, for both 10 mm and 15 mm distances, and there is no damage or burning on the tissues' surfaces. Varnagiris et al. developed a new technique, based on the combination of simultaneous non-thermal air plasma treatment with Mg nanoparticles deposition processes, applying to Mung bean seeds to enhance their quality [7]. This stimulates new chemical bond formations on the seed's surface, leading to a shift in surface characteristics, from hydrophobic to hydrophilic and, in turn, improving water uptake. They reported around two times better germination after 30 min of the plasma treatment, compared to the initial Mung bean seeds. At a more applied level, Attri and coworkers discussed plasma for agricultural applications, from laboratory to farm [8]. They concluded that the direct plasma treatments, working at low/medium/atmospheric pressure, and plasma-treated water (PTW) treatments changed the physical and biochemical properties of the seeds, and emphasized doing the real field experiments with the plasma-treated seeds to make them societally useful. They finally addressed the possibility of using plasma in the actual agricultural field and the prospects of this technology.

The above papers demonstrate the importance of the area of plasma diagnostics and applications, ranging from the formulation of fundamental theory to practical applications. Although the basic principles of plasma in the diagnostics and applications are well understood, the articles address outstanding challenges related to plasma in different areas, in terms of both applications and diagnostic perspectives, and much remains to be explored in the future. With the enormous variety and number of applications currently under development, we feel confident about the longevity and future of this subject.

We thank all the contributors and the Editor-in-Chief, Giancarlo Cravotto, for their enthusiastic support of this Special Issue, as well as the editorial staff of Processes for their efforts.

Funding: There is no funding support.

Conflicts of Interest: The authors declare no conflict of interest.

References

1. Wang, A.; Shang, J.; Wang, Q.; Wang, K. Effects of Cowl-Induced Expansion on the Wave Complex Induced by Oblique Detonation Wave Reflection. *Processes* **2021**, *9*, 1215. [CrossRef]
2. Xie, W.; Luo, Z.; Zhou, Y.; Wang, L.; Peng, W.; Gao, T. Experimental study on ramp shock wave control in Ma3 supersonic flow using two-electrode SparkJet actuator. *Processes* **2020**, *8*, 1679. [CrossRef]
3. Chen, W.; Jin, D.; Cui, W.; Huang, S. Characteristics of gliding arc plasma and its application in swirl flame static instability control. *Processes* **2020**, *8*, 684. [CrossRef]
4. Xu, Y.; Wang, Y.; Xu, Y.; Li, M.; Hu, Z. Microscopic Characteristics and Properties of Fe-Based Amorphous Alloy Compound Reinforced WC-Co-Based Coating via Plasma Spray Welding. *Processes* **2021**, *9*, 6. [CrossRef]
5. Hwang, S.H.; Koga, K.; Hao, Y.; Attri, P.; Okumura, T.; Kamataki, K.; Itagaki, N.; Shiratani, M.; Oh, J.-S.; Takabayashi, S. Time of Flight Size Control of Carbon Nanoparticles Using Ar+ CH4 Multi-Hollow Discharge Plasma Chemical Vapor Deposition Method. *Processes* **2021**, *9*, 2. [CrossRef]
6. Chen, Z.; Obenchain, R.; Wirz, R.E. Tiny cold atmospheric plasma jet for biomedical applications. *Processes* **2021**, *9*, 249. [CrossRef]
7. Varnagiris, S.; Vilimaite, S.; Mikelionyte, I.; Urbonavicius, M.; Tuckute, S.; Milcius, D. The combination of simultaneous plasma treatment with mg nanoparticles deposition technique for better mung bean seeds germination. *Processes* **2020**, *8*, 1575. [CrossRef]
8. Attri, P.; Ishikawa, K.; Okumura, T.; Koga, K.; Shiratani, M. Plasma agriculture from laboratory to farm: A review. *Processes* **2020**, *8*, 1002. [CrossRef]

Article

Effects of Cowl-Induced Expansion on the Wave Complex Induced by Oblique Detonation Wave Reflection

Aifeng Wang [1], Jiahao Shang [2], Qiu Wang [2] and Kuanliang Wang [3,*]

[1] Aero Engine Academy of China, Aero Engine Corporation of China, Beijing 101304, China; wangaifeng06@sina.com
[2] State Key Laboratory of High Temperature Gas Dynamics, Institute of Mechanics, Chinese Academy of Sciences, Beijing 100190, China; shangjiahao@imech.ac.cn (J.S.); wangqiu@imech.ac.cn (Q.W.)
[3] School of Aerospace Engineering, Beijing Institute of Technology, Beijing 100081, China
* Correspondence: wangkuanliang@bit.edu.cn

Abstract: Oblique detonation wave (ODW) reflection on the upper wall leads to a sophisticated wave complex, whose stability is critical to the application of oblique detonation engines. The unstable wave complex characterized with a continuous moving Mach stem has been observed, but the corresponding re-stability adjusting method is still unclear so far. In this study, the cowl-induced expansion wave based on the model with an upper-side expansion wall is introduced, and the ODW dynamics have been analyzed using the reactive Euler equations with a two-step induction–reaction kinetic model. With the addition of a cowl-induced expansion wave, the re-stabilized Mach stem has been distinguished. This re-stability is determined by the weakened secondary reflection wave of lower wall, while the final location of Mach stem is not sensitive to the position of the expansion corner. The re-stabilized ODW structure is also basically irrelevant to the expansion angle, while it may shift to unstable due to the merging of subsonic zones. Transient phenomena for the unstable state have been also discussed, clarifying fine wave structures further.

Keywords: oblique detonation; asymmetric; nozzle; reflection

Citation: Wang, A.; Shang, J.; Wang, Q.; Wang, K. Effects of Cowl-Induced Expansion on the Wave Complex Induced by Oblique Detonation Wave Reflection. *Processes* **2021**, *9*, 1215. https://doi.org/10.3390/pr9071215

Academic Editors: Alessandro D' Adamo and Albert Ratner

Received: 7 May 2021
Accepted: 7 July 2021
Published: 15 July 2021

1. Introduction

In the air-breathing propulsion field, the oblique detonation engine (ODE) concept has attracted more and more attention in recent years [1–4]. The gaseous combination of electrons, ions, and neutral species produced by the fast chemical reactions of ODE is a plasma. By utilizing oblique detonation waves (ODWs) triggered by ramps, ODE could achieve high thermal efficiency and fast heat release. However, the ODWs are usually complicated, and many studies have been conducted aiming to ascertain the coupling relation of shock and heat release [5–7]. Based on several previous studies [8–10], the wave structures and instability are analyzed in depth. Lately, three types of wave structures in the initiation region are concluded [11,12], depending on different aerodynamic [11,13,14] and chemical dynamic parameters [12,15]. A semi-theoretical model has been proposed which could predict the ODW's main features from the viewpoint of compression wave convergence, which derives from the heat release inside the supersonic flow [16].

Previous studies [17–20] investigated the ODWs in free space, which means that only a ramp triggering the wave is introduced, without considering the upper solid boundary limiting the oblique wave surface extending downstream. Early studies on the ram accelerator (another type of instrument utilizing the ODW) have considered the effects of an upper solid boundary, e.g., [21,22]. For realistic engine application, the upper wall of the flow tunnel always induces a wave reflection, and the corresponding wave complex structures need to be clarified. A systemic study on the ODW reflection before an expansion corner has been performed to explore the variation of wave structure [23,24]. It is observed that there are two possible wave complexes, one is featured by a recirculation zone, and the other is featured by a Mach stem. The wave complex with Mach stem might introduce a thermal choking, leading to an unstable process, whose mechanism is attributed to the combination of small post-shock subsonic zones [25].

The cowl-induced expansion has been studied before, e.g., [26–28], but not with the upper wall-induced expansion collectively. In this study, cowl-induced expansion is introduced based on the model with only one-side expansion, further approaching the realistic engine flow. The two-side expansion configuration will actually produce an asymmetric nozzle, whose performance is key to the engine. It is found out that the cowl-induced expansion might re-stabilize some unstable ODWs, which should be considered in the engine design. Detailed wave interactions were analyzed with some transient phenomena, and effects of expansion corner position and angle are investigated to clarify the stable and unstable transition mechanism.

2. Numerical Methods

A physical model of the engine and its main computational parameters are shown in Figure 1. For the simplified combustor–nozzle model (Figure 1a), ODW is first reflected by the combustor upper wall and then affected by the expanding nozzle. For the corresponding geometric parameters (Figure 1b), this study does not investigate the nozzle effects, but ascertains how the two-side expansion is different from the one-side one studied before. L_w and $\theta_{w'}$ are the two controlling parameters of the lower-side expansion: L_w represents the distance between the deflection corner of lower wedge and the wedge tip; $\theta_{w'}$ is the angle of deflection lower wedge. The upper expanding wall is set as follows: L_d is the distance between the outward point and the original undisturbed ODW surface, and θ_d is the outward turning angle. Two other main parameters H and θ_w, denoting the entrance inflow height and the wedge angle, are set be fixed.

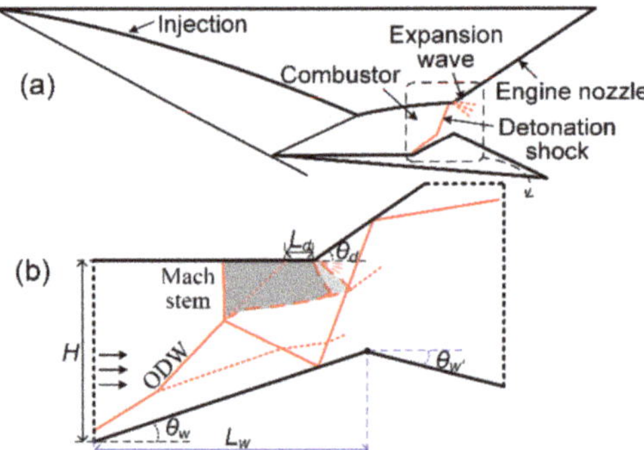

Figure 1. Schematic of oblique detonation engine (**a**) and wave complex (**b**).

Following our previous studies [23–25], the simulation is based on the Euler equations with a two-step kinetic model [29,30] by introducing two chemical variables: the induction reaction index ξ and the heat release reaction index λ:

$$\frac{\partial(\rho\xi)}{\partial t} + \frac{\partial(\rho u\xi)}{\partial x} + \frac{\partial(\rho v\xi)}{\partial y} = H(1-\xi)\rho k_I exp\left[E_I\left(\frac{1}{Ts} - \frac{1}{T}\right)\right] \quad (1)$$

$$\frac{\partial(\rho\xi)}{\partial t} + \frac{\partial(\rho u\xi)}{\partial x} + \frac{\partial(\rho v\xi)}{\partial y} = H(1-\xi)\rho k_I exp\left[E_I\left(\frac{1}{Ts} - \frac{1}{T}\right)\right], \quad (2)$$

$$\frac{\partial(\rho\lambda)}{\partial t} + \frac{\partial(\rho u\lambda)}{\partial x} + \frac{\partial(\rho v\lambda)}{\partial y} = [1-H(1-\xi)]\rho(1-\lambda)k_R exp\left[\left(-\frac{E_R}{T}\right)\right], \quad (3)$$

with the Heaviside step function:

$$H(1-\xi) = \begin{cases} 1, & \xi \le 1, \\ 0, & \xi > 1. \end{cases} \tag{4}$$

The equation of state includes the effects of heat release which is depended on the second step index λ:

$$e = \frac{p}{\rho(\gamma-1)} + \frac{1}{2}\left(u^2+v^2\right) - \lambda \tag{5}$$

The variables p, ρ, u, v, e, and Q are the pressure, density, x-direction velocity, y-direction velocity, specific total energy, and the amount of chemical heat release, respectively. All the variables have been non-dimensionlized by reference to the uniform unburned state as follows:

$$p = \frac{\tilde{p}}{p_0},\ \rho = \frac{\tilde{\rho}}{\rho_0},\ T = \frac{\tilde{T}}{T_0},\ u = \frac{\tilde{u}}{\sqrt{RT_0}},\ v = \frac{\tilde{v}}{\sqrt{RT_0}}, \tag{6}$$

The main chemical parameter used are set to be Q = 25, γ = 1.2, E_I = 4.0T_s and E_R = 1.0T_s, where T_s denotes the post-shock temperature of Chapman-Jouguet (C-J) detonation. To complete the kinetic model, the parameters k_I and k_R are necessary, and $k_I = -u_{vn}$ where u_{vn} is the particle velocity behind the shock front in the shock-fixed frame for the corresponding C-J detonation, whereby the induction length of the CJ detonation is fixed to unity, and the parameter k_R is set to be 1.0 to control the heat release rate process. These parameters do not correspond readily to any detailed reactants, but rather a generic model with modest heat release and activation energy.

The governing equations are solved by the Advection Upstream Splitting Method (AUSM)-type splitting with a third-order Monotone Upstream-Centered Schemes for Conservation Laws (MUSCL) approach. The third order Runge-Kutta algorithm [31] is used as the time-discretization scheme to achieve sufficient resolution for the simulations. For a higher resolution in capturing oblique shocks, the AUSMPW+ scheme is utilized by the introduction of a new numerical speed of sound and simplification for AUSMPW. The mesh scale of dx = 0.2 is used and verified by refining the grids, and the CFL number is 0.8.

The main parameters are listed in Table 1. Based on the parameter values set above, the typical ODW structures can be obtained with the inflow M_0 = 6.5–7.5, which is suitable for the engine application with high altitude. On the geometric parameters, most are fixed to be constant, such as H = 150, θ_w = 25° and θ_d = 45°, while L_w and $\theta_{w'}$ are variable as the bifurcation parameters to investigate the effects of the lower expansion corner.

Table 1. Simulation parameters used in the following cases.

Q	25
γ	1.2
E_I	4.0T_s
E_R	1.0T_s
M_0	6.5, 7.0, 7.5
θ_w	25°
θ_d	45°
H	150

3. Results and Discussion

First, the wedge-induced ODWs are simulated in the free space and the upper expansion combustor, respectively. As shown in Figure 2a, the original ODW without upper wall is a typical structure with a smooth transition from an oblique shock wave (OSW) to an ODW. When we consider the upper wall, which deflects outward after the original surface, a Mach reflection zone arises. Between the Mach stem and the expansion wave, there is a subsonic zone observed, as shown by white curve in Figure 2b. At the connection point of the ODW surface and Mach stem, a transverse shock extends downstream and reflects on the wedge. From the connection point, a slip line extends downstream which becomes unstable and induces vortex. Moreover, there is an interaction of slip line and second reflection of the wedge downstream.

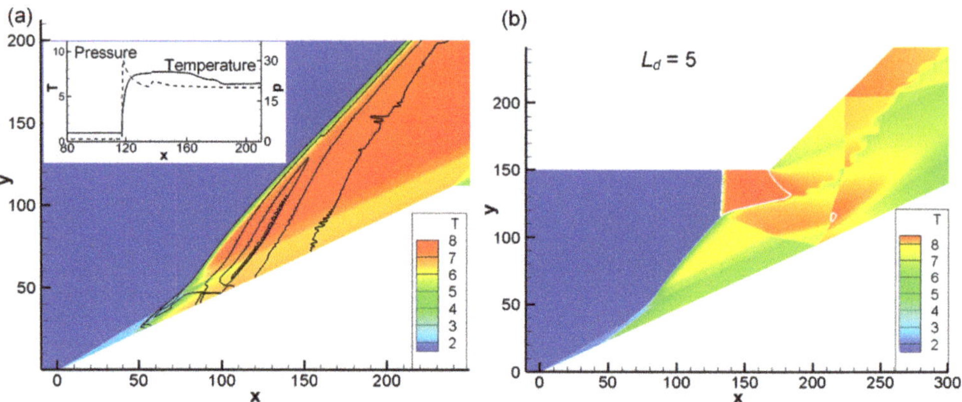

Figure 2. ODW temperature fields for M_0 =7.0, (**a**) without upper wall; (**b**) with upper wall and L_d = 5 (white curve denotes the sonic line).

Based on the stable wave complex above, an unstable one arises by increasing the distance between the deflected point and the original ODW surface (L_d) as shown in Figure 3.

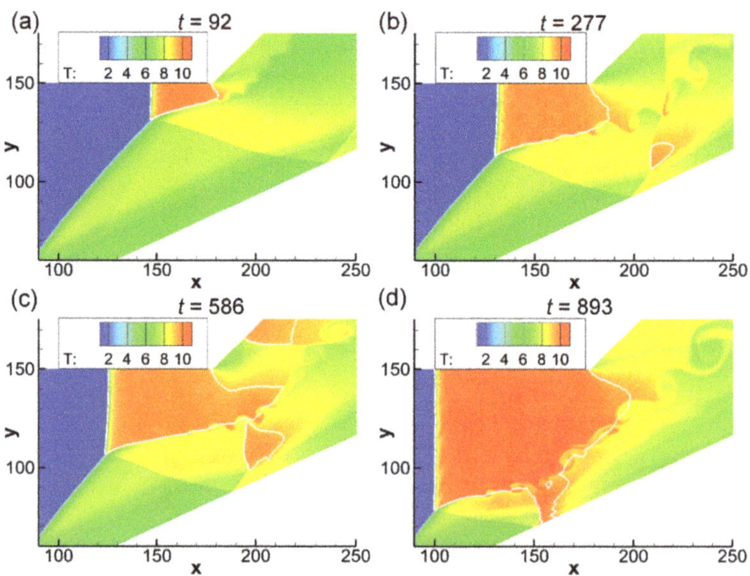

Figure 3. Unstable ODW evolution for M_0 =7.0, L_d = 15 at different instants, (**a**) t = 92; (**b**) t = 277; (**c**) t = 586; (**d**) t = 893.

Four sub-frames are plotted to display the evolution of unstable structures. As denoted by the time instant, the Mach stem is generated at the early stage and then travels upstream that make a similar wave complex like the above stable case. Nevertheless, the flow field then is not stationary, leading to the variation of wave locations. The second reflection of the wedge induced a new subsonic zone and gradually connected the rear boundary of the subsonic zone which also is extending downstream. The connection leads to the combination of two subsonic zones, and the wave complex becomes unstable that triggers the overall thermal choking of flow fields.

To suppress the process of thermal choking and stabilize the ODW surface, the cowl-induced expansion wave is introduced, which is specified by the corner location (L_w) and the deflection angle ($\theta_{w'}$). A series of flow fields are simulated by adjusting L_w in a decreasing manner with a fixed deflection angle of $\theta_{w'} = 0$, and we observe there is a boundary value that can stop the upstream movement of Mach stem. Figure 4 shows the typical re-stabilized structures corresponding to $M_0 = 7.0$, $L_d = 15$. As shown in Figure 4a, when the flow field becomes stable again, the reflection wave happens to act on the expansion corner, and the corresponding L_w value is about 180. The secondary reflection wave of the low wall is weakened so that it no longer affects the subsonic zone. By the effect of cowl-induced expansion wave, the downstream wave complex system becomes simple, and the heat release mainly occurs near the back of ODW and Mach stem surface. When the reflection wave acts before the expansion corner, i.e., L_w is set large than 180, an unstable evolution will still occur despite the existence of the expansion wave.

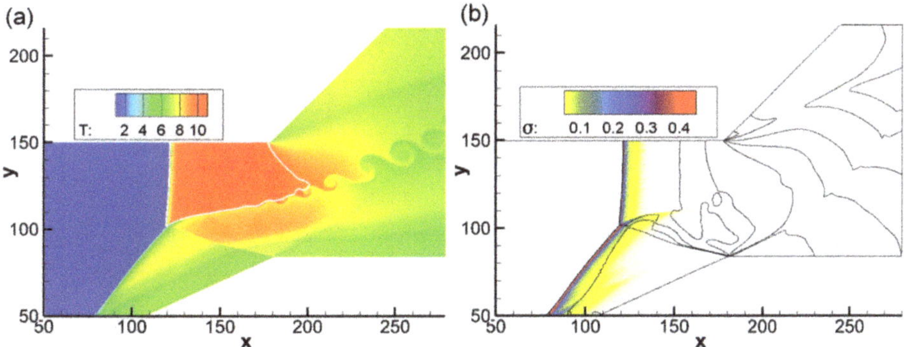

Figure 4. Re-stabilized ODW structure for M_0 =7.0, L_d = 15, L_w = 180 (**a**) temperature field with white curve denoting the sonic line; (**b**) heat release field with pressure contours.

The sensitivity of L_w for the flow field is analyzed by examining the Mach stem position at the same instant, and the results are shown in Figure 5. L_w is changed in a range of 160 to 200 with an interval of 5, and the instant is chosen when the Mach stem was about to lose stability for the first unstable case (L_w = 185). In this study, L_M is defined by the length of the post-stem subsonic zone, starting from the Mach stem to the expansion corner along the upper wall. With a large L_w, the Mach stem travels upstream and does not stop due to the thermal choking, so L_M approaches infinity for unstable cases. However, it could be observed that with a L_w small enough, L_M approaches about 57, and the length is not sensitive to L_w. This indicates that the Mach stem position is fixed in the same when the reflection wave acts after the expansion corner. The introduction of cowl-induced expansion effectively prevents the occurrence of thermal choking and its location will not change the stable ODW structure, which is a good thing for the combustor design.

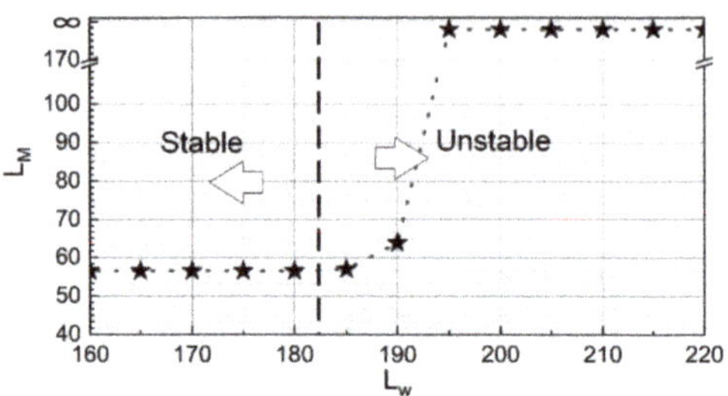

Figure 5. The relation of L_w-L_M for M_0 =7.0, L_d = 15.

The wave complex stability depends on resolving the triple point and the Mach stem. Therefore, the resolution studies were performed for the two typical cases of $L_d = 5$, $L_w = \infty$ and $L_d = 15$, $L_w = 180$ respectively. The two cases respectively represent the stability of the upper wall-induced expansion and the cowl-induced expansion, as shown in Figures 2b and 4. The wave structures on two grid size scales ($dx = 0.20$ and $dx = 0.15$) are compared in Figure 6, and to distinguish the effects of grid resolution for the triple point and the Mach stem, the, we quantitatively compare the temperature and pressure along $y = 125$ in Figure 7.

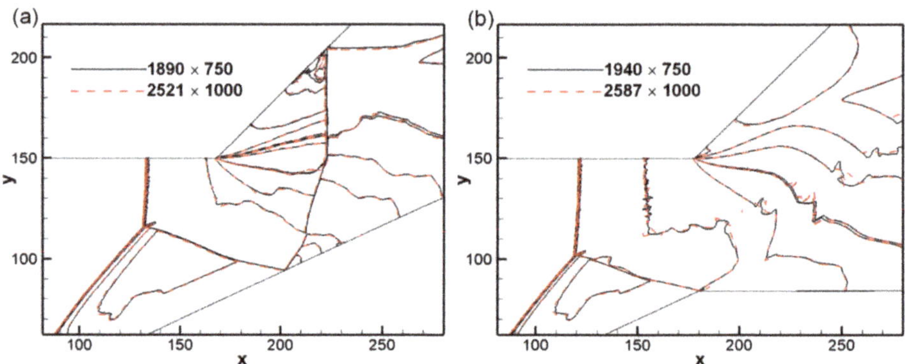

Figure 6. Temperature contours from different grids (**a**) M_0 =7.0, L_d = 5, $L_w = \infty$; (**b**) M_0 =7.0, L_d = 15, L_w = 180.

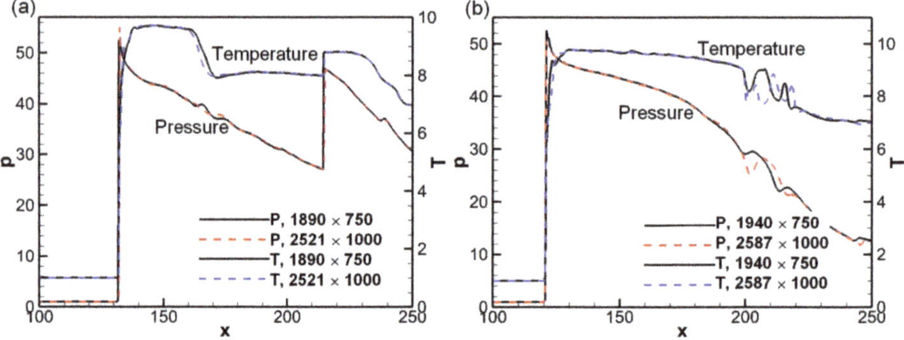

Figure 7. Pressure and temperature along certain lines of above two cases (**a**) M_0 =7.0, L_d = 5, $L_w = \infty$; (**b**) M_0 =7.0, L_d = 15, L_w = 180.

In Figures 6 and 7, the black solid lines show the results for the grid length scale $dx = 0.2$, and the dotted lines show those for $dx = 0.15$. For both cases, the triple point and the Mach stem positions keep in the same positions, and except for the slight discrepancy in the vortex position of the downstream slip line, the difference in the overall wave structure is almost negligible. Considering the vortex rolling should be attributed to the Kelvin–Helmholtz instability (KHI) of the slip line, this discrepancy does not affect the overall ODW dynamics. The curve of pressure and temperature in Figure 7 show again good agreement, therefore, the chosen grid size is sufficient to reveal the global structures for the purpose of this work.

The unstable ODWs for $M_0 = 6.5$ and $M_0 = 7.5$ are also investigated, and the results show a similar thermal choking evolution. As shown in Figure 8, the detonation surface at different time is outlined by the white lines. The Mach stem is first formed by the reflection of original ODW on upper wall and then moves upstream over time without stopping. Given $L_d = 1$ for $M_0 = 6.5$, an unstable process appears, but the corresponding L_d length for $M_0 = 7.5$ requires to be 30. It can be concluded that the instability characteristics are the same, though the L_d length at which instability occurs increases with the Mach number. The moving Mach stem can also be stopped by the adding of cowl-induced expansion, and Figure 9 shows the typical re-stabilized flow fields of the two Mach numbers. Compared with the stable ODW in Figure 4, i.e., the case of $M_0 = 7.0$ discussed before, both the aera and the temperature of subsonic zone increase with the Mach number, but the main flow features are basically unchanged that the secondary reflection wave of the wall cannot disturb the upstream subsonic zone.

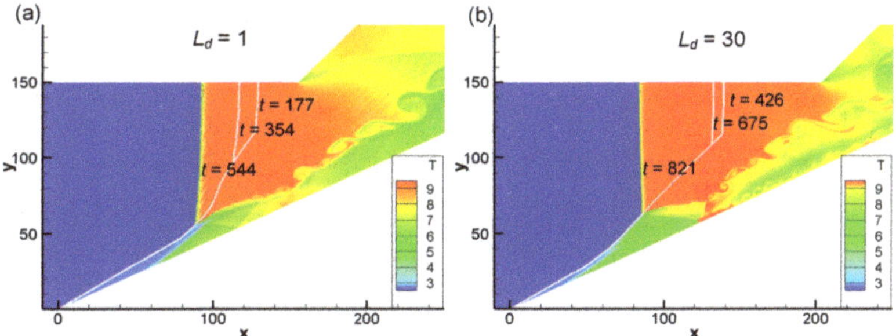

Figure 8. Unstable ODW evolutions (**a**) $M_0 = 6.5$, $L_d = 1$, $L_w = \infty$; (**b**) $M_0 = 7.5$, $L_d = 30$, $L_w = \infty$.

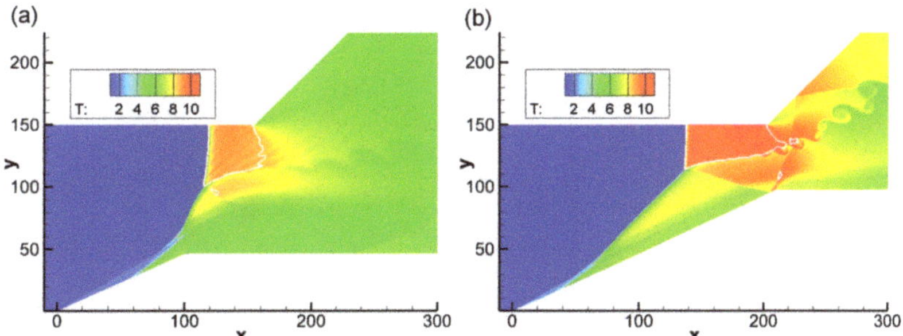

Figure 9. Re-stabilized ODW structures (**a**) $M_0 = 6.5$, $L_d = 1$, $L_w = 100$; (**b**) $M_0 = 7.5$, $L_d = 30$, $L_w = 210$.

The effect of lower deflection angle ($\theta_{w'}$) is examined based on the stable cases of $M_0 = 7.0$. $\theta_{w'}$ is changed in a range of $-20°$ to $20°$ with an interval of $5°$, and the unstable ODW arises again when $\theta_{w'}$ increases to $20°$ in Figure 10b. For the stable ODW complex, it can be seen from the comparison of Figures 4 and 10a that the secondary reflection of the lower wall weakens due to the extending of expansion angle. The downstream wave complex remains fixed, therefore the change of $\theta_{w'}$ will not change the Mach stem position and the corresponding structure. As the lower expansion wall changes upward, the secondary reflection wave strengthens and deflects upstream. If the wave acts on the subsonic zone, instability will happen. To analyze the unstable ODW complex, the sonic lines at different time are shown in Figure 10b. For the unstable ODW complex, as denoted by the time instant, the early-stage flow field at t = 570 has a similar wave complex like the steady case of Figure 10a. Nevertheless, the Mach stem then is not stationary, and the subsonic zone extends downstream. The secondary reflection wave also induces a subsonic zone and connects with the upper one at t = 1880. The connection leads to the combination and expanding of two zones, and travels upstream continuously, triggering the unstable wave complex.

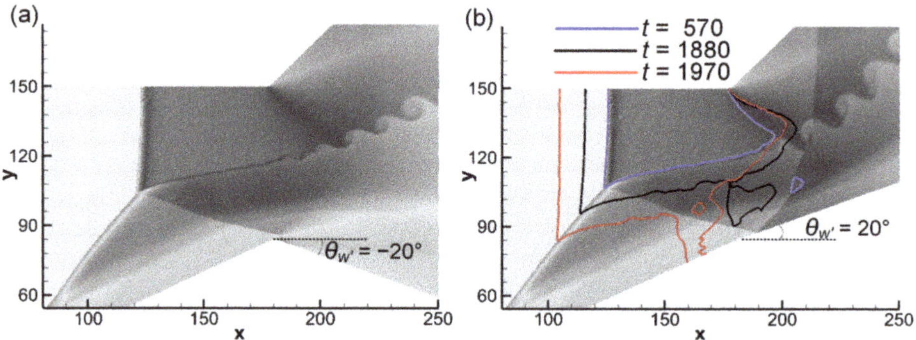

Figure 10. Effects of lower deflection angles for M_0 = 7.0, L_d = 15, L_w = 180, (**a**) $\theta_{w'}$ = −20°; (**b**) $\theta_{w'}$ = 20°.

As mentioned above in Figure 5, the ODW structure becomes unstable when L_w increases to 185 with the horizontal deflection wall ($\theta_{w'}$ = 0). We try to get the stable ODW structure of L_w = 185 by expanding $\theta_{w'}$ to −30° (Figure 11), but the trail does not success finally. As shown by the flow dynamics in Figure 9, the combination of the two subsonic zones also happens, so the expanding of expansion angle cannot suppress the process of thermal choking. From the viewpoint of convergent-divergent flow, these cases could be explained by the hypothesis of one-dimensional isentropic: the convergence divergence ratios not yet reach the limit of the thermal choking. Therefore, the unstable evolution is controlled by the subsonic zone and the behavior of secondary reflection wave.

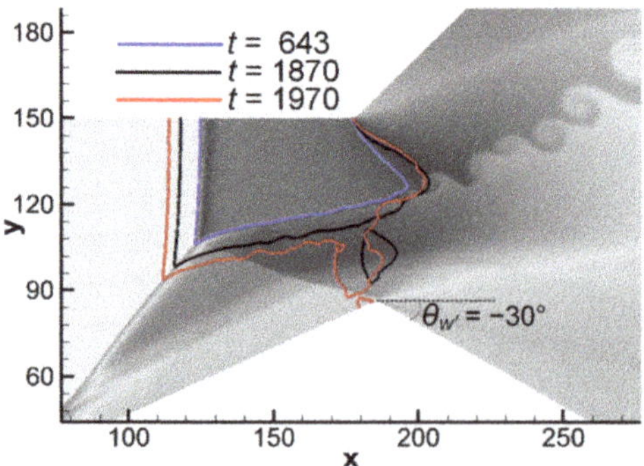

Figure 11. Unstable ODW for M_0 =7.0, L_d = 15, L_w = 185.

4. Conclusions

The wave dynamic features and the unstability rules of the cowl-induced expansion for the ODW reflection before expansion corner are investigated in this study. For the ODW reflection of only upper-side expansion, a Mach stem and the corresponding subsonic zone are produced, and the unstable mode of the wave complex has been observed. By the introduction of cowl-induced expansion, the numerical results show that the upstream movement of Mach stem can be re-stabilized for the unstable structure. The critical value of L_w is found, which is determined by the secondary reflection wave of the lower wall, while the position of stable Mach stem is not sensitive to it. By adjusting the degree of the cowl-induced expansion angle, the re-stabilized ODW shifts to unstable mode due to the influence of secondary reflection wave. However, the structure characteristics of re-stabilized ODW and the unstable process are not related with expansion angle. The unstable structures bring challenges to the application of ODW-based engines, but the numerical results suggest a feasible adjustment method and the corresponding transient phenomena deserve more attention in future work.

Author Contributions: Conceptualization, methodology, A.W. and J.S.; formal analysis, investigation, Q.W.; project administration, writing, K.W. All authors have read and agreed to the published version of the manuscript.

Funding: This research was supported by the National Natural Science Foundation of China (NSFC) (Nos. 12002041).

Data Availability Statement: Data is contained within the article.

Conflicts of Interest: The authors declare no conflict of interest.

References

1. Ashford, S.A.; Emanuel, G. Oblique detonation wave engine performance prediction. *J. Propul. Power* **1996**, *12*, 322–327. [CrossRef]
2. Sislian, J.P.; Schirmer, H.; Dudebout, R.; Schumacher, J. Propulsive performance of hypersonic oblique detonation wave and shock-induced combustion ramjets. *J. Propul. Power* **2001**, *17*, 599–604. [CrossRef]
3. Urzay, J. Supersonic combustion in air-breathing propulsion systems for hypersonic flight. *Annu. Rev. Fluid Mech.* **2018**, *50*, 593–627. [CrossRef]
4. Vanyai, T.; Bricalli, M.; Brieschenk, S.; Boyce, R.R. Scramjet performance for ideal combustion processes. *Aerosp. Sci. Technol.* **2018**, *75*, 215–226. [CrossRef]
5. Choi, J.Y.; Kim, D.W.; Jeung, I.S.; Ma, F.; Yang, V. Cell-like structure of unstable oblique detonation wave from high-resolution numerical simulation. *Proc. Combust. Inst.* **2007**, *31*, 2473–2480. [CrossRef]
6. Teng, H.H.; Jiang, Z.L. On the transition pattern of the oblique detonation structure. *J. Fluid Mech.* **2012**, *713*, 659–669. [CrossRef]
7. Braun, E.M.; Lu, F.K.; Wilson, D.R.; Camberos, J.A. Airbreathing rotating detonation wave engine cycle analysis. *Aerosp. Sci. Technol.* **2013**, *27*, 201–208. [CrossRef]
8. Viguier, C.; Figueira Da Silva, L.F.; Desbordes, D.; Deshaies, B. Onset of oblique detonation waves: Comparison between experimental and numerical results for hydrogen-air mixtures. *Symp. (Int.) Combust.* **1996**, *26*, 3023–3031. [CrossRef]
9. Verreault, J.; Higgins, A.J.; Stowe, R.A. Formation of transverse waves in oblique detonations. *Proc. Combust. Inst.* **2013**, *34*, 1913–1920. [CrossRef]
10. Yang, P.; Ng, H.D.; Teng, H. Unsteady dynamics of wedge-induced oblique detonations under periodic inflows. *Phys. Fluids* **2021**, *33*, 016107. [CrossRef]
11. Bian, J.; Zhou, L.; Teng, H. Structural and thermal analysis on oblique detonation influenced by different forebody compressions in hydrogen-air mixtures. *Fuel* **2021**, *286*, 119458. [CrossRef]
12. Yang, P.; Teng, H.H.; Jiang, Z.L.; Ng, H.D. Effects of inflow Mach number on oblique detonation initiation with a two-step induction-reaction kinetic model. *Combust. Flame* **2018**, *193*, 246–256. [CrossRef]
13. Yu, M.; Miao, S. Initiation characteristics of wedge-induced oblique detonation waves in turbulence flows. *Acta Astronaut.* **2018**, *147*, 195–204. [CrossRef]
14. Teng, H.; Bian, J.; Zhou, L.; Zhang, Y. A numerical investigation of oblique detonation waves in hydrogen-air mixtures at low mach numbers. *Int. J. Hydrogen Energy* **2021**, *46*, 10984–10994. [CrossRef]
15. Iwata, K.; Nakaya, S.; Tsue, M. Wedge-stabilized oblique detonation in an inhomogeneous hydrogen-air mixture. *Proc. Combust. Inst.* **2017**, *36*, 2761–2769. [CrossRef]
16. Teng, H.; Tian, C.; Zhang, Y.; Zhou, L.; Ng, H.D. Morphology of oblique detonation waves in a stoichiometric hydrogen–air mixture. *J. Fluid Mech.* **2021**, *913*, A1. [CrossRef]

17. Pratt, D.T.; Humphrey, J.W.; Glenn, D.E. Morphology of standing oblique detonation waves. *J. Propul. Power* **1991**, *7*, 837–845. [CrossRef]
18. Li, C.; Kailasanath, K.; Oran, E.S. Detonation structures behind oblique shocks. *Phys. Fluids* **1994**, *6*, 1600–1611. [CrossRef]
19. Teng, H.H.; Jiang, Z.L.; Ng, H.D. Numerical study on unstable surfaces of oblique detonations. *J. Fluid Mech.* **2014**, *744*, 111–128. [CrossRef]
20. Teng, H.; Ng, H.D.; Kang, L.; Luo, C.; Jiang, Z. Evolution of cellular structures on oblique detonation surfaces. *Combust. Flame* **2015**, *162*, 470–477. [CrossRef]
21. Schultz, E.; Knowlen, C.; Bruckner, A.P. Starting Envelope of the Subdetonative Ram Accelerator. *J. Propul. Power* **2000**, *16*, 1040–1052. [CrossRef]
22. Bachman, C.L.; Goodwin, G.B. Ignition criteria and the effect of boundary layers on wedge-stabilized oblique detonation waves. *Combust. Flame* **2021**, *223*, 271–283. [CrossRef]
23. Wang, K.; Teng, H.; Yang, P.; Ng, H.D. Numerical investigation of flow structures resulting from the interaction between an oblique detonation wave and an upper expansion corner. *J. Fluid Mech.* **2020**, *903*, A28. [CrossRef]
24. Wang, K.; Zhang, Z.; Yang, P.; Teng, H. Numerical study on reflection of an oblique detonation wave on an outward turning wall. *Phys. Fluids* **2020**, *32*, 046101.
25. Wang, K.; Yang, P.; Teng, H. Steadiness of wave complex induced by oblique detonation wave reflection before an expansion corner. *Aerosp. Sci. Technol.* **2021**, *112*, 106592. [CrossRef]
26. Bhattrai, B.; Tang, H. Formation of near-Chapman–Jouguet oblique detonation wave over a dual-angle ramp. *Aerosp. Sci. Technol.* **2017**, *63*, 1–8. [CrossRef]
27. Choi, J.Y.; Shin, E.J.; Jeung, I.S. Unstable combustion induced by oblique shock waves at the non-attaching condition of the oblique detonation wave. *Proc. Combust. Inst.* **2009**, *32*, 2387–2396. [CrossRef]
28. Xiang, G.X.; Gao, X.; Tang, W.J.; Jie, X.Z.; Huang, X. Numerical study on transition structures of oblique detonations with expansion wave from finite-length cowl. *Phys. Fluids* **2020**, *32*, 056108. [CrossRef]
29. Korobeinikov, V.P.; Levin, V.A.; Markov, V.V.; Chernyi, G.G. Propagation of blast waves in a combustible gas. *Astronaut. Acta* **1972**, *17*, 529–537.
30. Ng, H.D.; Radulescu, M.I.; Higgins, A.J.; Nikiforakis, N.; Lee, J.H.S. Numerical investigation of the instability for one-dimensional Chapman-Jouguet detonations with chain-branching kinetics. *Combust. Theory Model* **2005**, *9*, 385–401. [CrossRef]
31. Kim, K.H.; Kim, C.; Rho, O.-H. Methods for the Accurate Computations of Hypersonic Flows: I. AUSMPW+Scheme. *J. Comput. Phys.* **2001**, *174*, 38–80. [CrossRef]

Article

Experimental Study on Ramp Shock Wave Control in Ma3 Supersonic Flow Using Two-Electrode SparkJet Actuator

Wei Xie, Zhenbing Luo *, Yan Zhou *, Lin Wang, Wenqiang Peng and Tianxiang Gao

Department of Applied Mechanics, School of Aerospace Sciences, National University of Defense Technology, Changsha 410073, China; xxiewei@nudt.edu.cn (W.X.); wanglin-2007@126.com (L.W.); plxhaz@126.com (W.P.); gaotxnudta@163.com (T.G.)

* Correspondence: luozhenbing@163.com (Z.L.); bjlgzy@163.com (Y.Z.)

Received: 24 November 2020; Accepted: 17 December 2020; Published: 19 December 2020

Abstract: The control of a shock wave produced by a ramp (ramp shock) in Ma3 supersonic flow using a two-electrode SparkJet (SPJ) actuator in a single-pulse mode is studied experimentally. Except for schlieren images of the interaction process of SPJ with the flow field, a dynamic pressure measurement method is also used in the analysis of shock wave control. In a typical experimental case, under the control of single-pulsed SPJ, the characteristic of ramp shock changes from "short-term local upstream motion" in the initial stage to "long-term whole downstream motion" in the later stage. The angle and position of the ramp shock changes significantly in the whole control process. In addition, the dynamic pressure measurement result shows that the ramp pressure is reduced by a maximum of 79% compared to that in the base flow field, which indicates that the ramp shock is significantly weakened by SPJ. The effects of some parameters on the control effect of SPJ on the ramp shock are investigated and analyzed in detail. The increase in discharge capacitance helps to improve the control effect of SPJ on the ramp shock. However, the control effect of the SPJ actuator with medium exit diameter is better than that with a too small or too large one. In addition, when the SPJ exit is located in the separation zone and outside, the change in the ramp shock shows significant differences, but the control effect in the case of medium ramp distance is better when the SPJ exit is located outside the separation zone.

Keywords: SparkJet actuator; shock wave control; supersonic flow; schlieren images; dynamic pressure measurement

---

1. Introduction

Gas discharge plasma technology has been developing rapidly and applied in many fields by researchers including medicine, material science, food science, aerospace science and so on in recent years. Its application in aerospace is very promising, mainly involving deicing [1–3], engine ignition and combustion [4,5], flow control [6–9] and so on. Plasma flow control refers to a method to apply effective disturbance to the flow field through the movement of plasma or the pressure, temperature, etc., changes due to plasma under the action of electromagnetic field force or gas discharge. It has wide application prospects in improving vehicles/engine aerodynamic characteristics, of which shock wave control is an important part.

Shock wave is a unique aerodynamic phenomenon in the era of supersonic flight. Effective control of the shock wave can improve the performance of supersonic flight in many aspects, such as reducing drag and heat, reducing the sonic boom of vehicle, generating thrust vector and regulating the capture flow rate and total pressure loss of ram inlet and so on. In order to control shock waves, researchers have adopted various passive/active flow control methods. The gas discharge plasma control method is a new kind of active flow control technology developed rapidly in recent years. It has many advantages, such as no moving parts or fluid supply device, fast response, wide working frequency band, etc. The SparkJet (SPJ) actuator, also called plasma synthetic jet actuator, is a new kind of plasma flow control method proposed by Grossman in 2003 [10]. It has very simple structures composed of an insulated cavity with a small exit and a pair of electrodes. A high voltage is applied between the two electrodes for the breakdown of cavity gas, and the gas in the small, insulated cavity is rapidly heated and pressurized. As a result, high-temperature, high-speed SPJ and a strong compression wave (also called "precursor shock" or "blast wave") [11,12] are formed at the exit of SPJ actuator (SPJ exit), which can then be used for flow control. The authors have studied the characterization of SPJ in quiescent air [12]. For a typical SPJ, schlieren image results showed that SPJ took on a typical mushroom shaped jet structure and gradually became a fully developed continuous turbulence. A strong blast shock wave and some weak reflected waves are formed in front of SPJ. Through calculation, with input energy of about 10.5 J, the velocity of "blast shock wave" is maintained at about 350 m/s, that is, the local sound velocity, so the blast shock wave is actually a strong compression wave propagating at the sound velocity. While the velocity of SPJ tends to decrease with multiple peaks over time, of which the peak velocity is about 300 m/s [12]. After SPJ is ejected, SPJ actuator will aspirate the air in the environment to prepare for the next discharge. The whole process takes only a few hundred microseconds. The SPJ actuator has been a hot research topic in plasma flow control field for the past 17 years [13–18]. On the one hand, the researchers conducted a detailed parametric study of SPJ actuator [19–23]. On the other hand, a large number of studies focused on the working characteristic [24–26], efficiency improvement [27,28] and structure optimization [29–31] of SPJ actuator. In addition, SPJ actuator has also been used for aerodynamic control [32,33], mixing enhancement [34] and separation control [35–38].

At present, the SPJ actuator has been preliminarily applied in supersonic flow field shock wave control. Cybyk et al. [39] proved that PSJ can penetrate the boundary layer of the supersonic flow field (Ma = 3) through numerical simulation, which causes the transition of the transverse main flow boundary layer. This is the first time to verify the control authority of PSJ actuator in supersonic flow. Phase-locked schlieren imaging was used by Narayanaswamy et al. [40] to estimate the strength of PSJ in Mach 3 transverse flow, and the penetration distance measured is 1.5 boundary-layer thicknesses. In his later publications, he used SPJ to control shock wave boundary layer disturbance and achieved good results [41,42]. Wang et al. [43] studied how the transverse SPJ interacted with the shock induced by the 24° ramp in a supersonic flow with Mach number 2. The schlieren images showed that the shock was significantly modified by SPJ with an upstream motion and a reduced angle. Huang et al. [44] found that a shock-on-shock interaction occurred when the SPJ shock intersected with the shock induced by a 20° compression ramp in Mach 2 flow. Research by Zhou et al. [45,46] proved that SPJ can significantly weaken the ramp shock both in supersonic and hypersonic flow.

Overall, in the current study of shock wave control using SPJ, schlieren images are mainly used for qualitative observation of the change in the controlled shock. Although to a certain extent, this method can explain the attenuation of the shock wave, but it is relatively rough and lacking in a more credible and quantitative conclusion. In addition, detailed parameter studies of shock wave control using SPJ are also lacking. Therefore, in this paper, the control of shock wave produced by a ramp (ramp shock) in Ma3 supersonic flow using a single-pulsed SPJ actuator is studied experimentally both using schlieren images and dynamic pressure measurement method. The effects of different parameters on the control effect of SPJ on the shock wave are also studied.

2. Experimental Setup

2.1. Supersonic Wind Tunnel

This experiment was carried out in KD-2 supersonic wind tunnel of National University of Defense Technology. As shown in Figure 1, the wind tunnel mainly composites of transition section, stable section, nozzle, experimental section, expansion section and vacuum tank. The experimental section measures 200 mm (width) × 200 mm (height) × 400 mm (length) and has large optical windows for flow visualization in all four directions. The experimental results show that the experimental section has a good laminar flow. In detail, the maximum fluctuation error of the transient velocity is less than 1%, and the absolute error of the nozzle outlet Mach number distribution is less than 2%. The basic flow parameters in the experimental section in this experiment are shown in Table 1. [47].

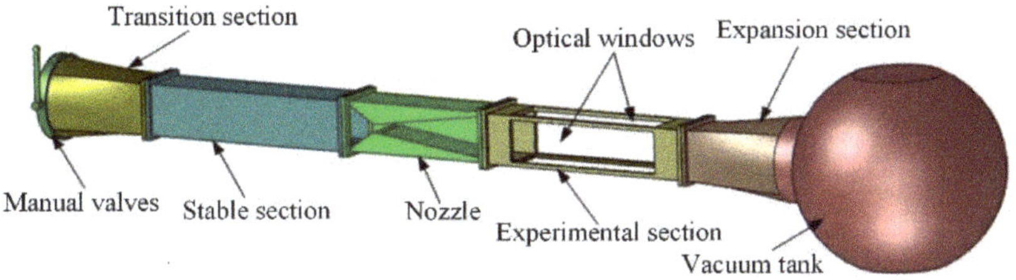

Figure 1. Schematic diagram of KD-2 supersonic wind tunnel.

Table 1. Flow parameters in the experimental section.

Parameters	Value	Unit
Mach number	3	-
Velocity	622.5	m/s
Sound velocity	207.5	m/s
Total temperature	300	K
Static temperature	107.1	K
Total pressure	101	kPa
Static pressure	2.8	kPa
Density	0.0983	kg/m^3
Viscosity coefficient	7.43×10^{-6}	Ns/m^2
Unit Reynolds number	7.49×10^{6}	1/m
Running time	>20	s

2.2. Experimental Model

The experimental model for shock wave control using transverse SPJ is shown in Figure 2a. The center plate was installed in the experimental section by the four brackets and was calibrated with a level meter during installation. The ramp was installed on the center plate to generate ramp shock. The width and height of the ramp are, respectively, 15 and 20 mm. Angle of the ramp shown in Figure 2a is 60°, and the ramp distance (the distance between SPJ exit and the root of the ramp) is 50 mm. A pressure measuring hole is on the ramp surface at a height of 15 mm from the center plate and is connected with the dynamic pressure sensor. The SPJ actuator was installed under the center plate and was mounted by a fixed seat. The structure of SPJ actuator is shown in Figure 2b. The discharge cavity is cylindrical, with radius and height of 5.4 and 10.8 mm, respectively, and the volume is 1000 mm^3. The cover plate can be tightly connected with the actuator mounting hole on the center plate. The cover plate and the actuator shell were fixed and sealed by silicone rubber to form a discharge cavity. Two tungsten electrodes were inserted into the discharge cavity for gas breakdown. In order to avoid leakage of electricity, an actuator sleeve was added outside the actuator shell to form the electrode–conductor connection groove, which was filled with insulating sealant. The agreed coordinate system in the experiment is shown in Figure 2, the coordinate system origin is located at the center of SPJ exit. When the actuator was in operation, a capacitor was connected in parallel at both ends to speed up discharge. The voltage used for discharge came from a high-voltage pulse power supply, which can supply voltage up to 10 kV.

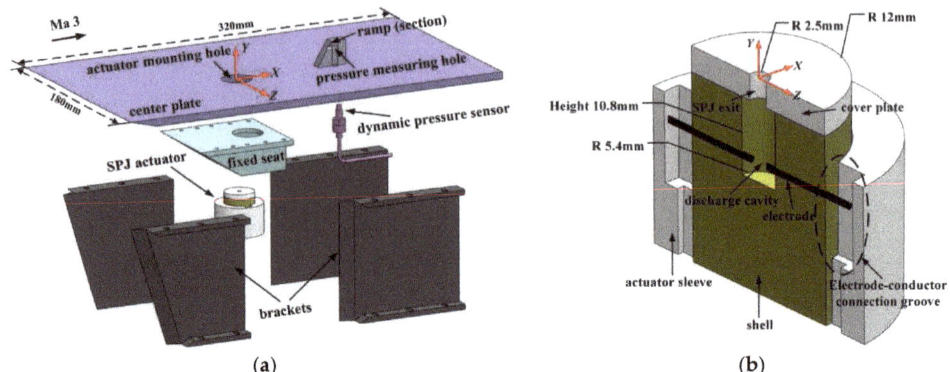

Figure 2. Schematic diagram of experimental model; (**a**) experimental model installation; (**b**) SparkJet (SPJ) actuator section view.

2.3. Measurement System

The voltage was measured using a high-voltage probe (Tektronix P6015A, 75 MHz, 0–20 kV). The voltage signals were recorded by an oscilloscope (Tektronix DPO3014, 100 MHz, 2.5 Gs/s).

The pressure change in the supersonic flow field ramp wall under the control of SparkJet is extremely fast, and the time is very short, so the pressure needs to be measured by a dynamic pressure sensor with high response frequency. In this experiment, kulite XTL-190, with range of 170 kPa (absolute pressure) and the inherent frequency of 240 kHz, was adopted. The kulite sensor is very small, so it is easier to be mounted into the ramp. In addition, the kulite sensor has high sensitivity and can measure dynamic pressure and static pressure of the flow field at the same time.

Interaction processes between SPJ and ramp shock were acquired using high-speed schlieren imaging. A standard Z-type schlieren setup was used in this experiment. The frame rate of the camera was set to 50 kHz, so the time interval between two images is 20 μs.

2.4. Experimental Cases Set

Case1 is the reference case, with discharge capacitance of 640 nF, SPJ actuator exit diameter of 5 mm, ramp distance of 50 mm and ramp angle of 60°. To study the effects of different parameters on the control of ramp shock using SPJ actuator, 10 cases are set as shown in Table 2. Case1, case2 and case3 are set for comparing the effect of discharge capacitance. Case4, case5 and case6 are set for comparing the effect of actuator exit diameter. Case1, case5, case7, case8, case9 and case10 are set for comparing the effect of ramp distance. Any other parameters that are not listed in Table 2 are the same in all the 10 cases, for example the volume of the discharge cavity remains 1000 mm^3. The SPJ actuator operation was in a single-pulse mode in all the 10 cases in this paper.

Table 2. Experimental cases set.

Parameters	Discharge Capacitance	Exit Diameter	Ramp Distance	Ramp Angle
Case1	640 nF	5 mm	50 mm	60°
Case2	320 nF	5 mm	50 mm	60°
Case3	80 nF	5 mm	50 mm	60°
Case4	640 nF	1.5 mm	75 mm	90°
Case5	640 nF	5 mm	75 mm	90°
Case6	640 nF	11 mm	75 mm	90°
Case7	640 nF	5 mm	15 mm	90°
Case8	640 nF	5 mm	30 mm	60°
Case9	640 nF	5 mm	70 mm	60°
Case10	640 nF	5 mm	90 mm	60°

3. Experimental Results

3.1. Shock Wave Attenuation and Elimination Characteristics

Figure 3 shows the interaction process between SPJ and the supersonic crossflow near the ramp in case1 during a single-pulse operation using schlieren images. At 0 μs, the discharge in the cavity begins, but SPJ is not produced yet, so it can be regarded as the base flow field. Because of the existence of center plate leading edge and the gap between the wind tunnel and the optical windows, two weak Mach waves are produced in the experimental section. Laminar flow field separates at the ramp root, with a separation zone and a weak separation shock upstream the ramp [47]. The ramp angle is greater than the critical compression angle in Ma3 supersonic flow, so a detached shock wave is formed upstream of the ramp, which is called ramp shock and regarded as the control object of SPJ in this paper. For the convenience of comparison, the position of ramp shock in the base flow field is marked with white dotted lines in Figure 3.

At 20 μs, due to the rapid heating effect in the discharge cavity, SPJ is ejected from the SPJ exit and begins to interact with the supersonic crossflow. At the same time, a weak ellipse-shaped compression wave called "SPJ shock" in this paper is formed upstream the SPJ. From 20 to 80 μs, SPJ and SPJ shock expands further in the flow and normal direction, but faster in the flow direction. From 80 to 120 μs, SPJ shock interacts with ramp shock and part of SPJ shock goes through ramp shock, but there is no obvious change for the ramp shock. The part of SPJ shock downstream the ramp shock is too weak to be seen after 140 μs. Then from 120 to 160 μs, SPJ begins to interact with ramp shock. With control of SPJ and SPJ shock, ramp shock is first characterized by "short-term local upstream motion". During this period of time, the effect of high-temperature SPJ is dominant. The local flow temperature and sound velocity rise due to the heating of the high-temperature SPJ, resulting in the "blocking effect", which is equivalent to expanding the shape of the ramp. Therefore, the ramp shock moves upstream under the blocking effect. However, as the high-temperature SPJ quickly moves downstream, it cannot exert further effects on the ramp shock. At 140 μs, the upstream motion of the ramp shock is the most significant, and then at 160 μs, the ramp shock quickly recovers roughly to the position in the base flow field.

Then from 200 to 1220 μs, ramp shock is characterized by "long-term whole downstream motion". The reasons may be explained as follows. On the one hand, the increase in the height of the SPJ shock causes stronger and wider disturbance to the ramp shock; on the other hand, it may be caused by the oscillation of the position of the ramp shock after the "short-term local upstream motion" in the previous stage. Although the ramp shock moves downstream as a whole finally, there is a process that the downstream-moving part gradually expands from the near-wall part to the far-wall part. For example, at 200 and 240 μs, the downstream-moving part is located below 53 and 64.5 mm, respectively, from the center plate wall. After 340 μs, the ramp shock in the whole observation area moves downstream. At 420 μs, the angle of the tail of the ramp shock reaches the minimum value (about 26.4°), which can be considered that the "long-term whole downstream motion" of the ramp shock reaches the maximum. Then, the ramp shock begins to recover gradually towards the position in the base flow field, but the recovery speed is very slow and gradually decelerates. Taking the angle of the tail of the ramp shock as a reference, as shown in Figure 4, the average recovery speed in the early stage (420–700 μs) is about 0.75° per 100 μs, and in the later stage (700–1220 μs) is about 0.29° per 100 μs. At 1220 μs, the ramp shock basically returns to the base flow state, which is the end of a control cycle of SPJ actuator.

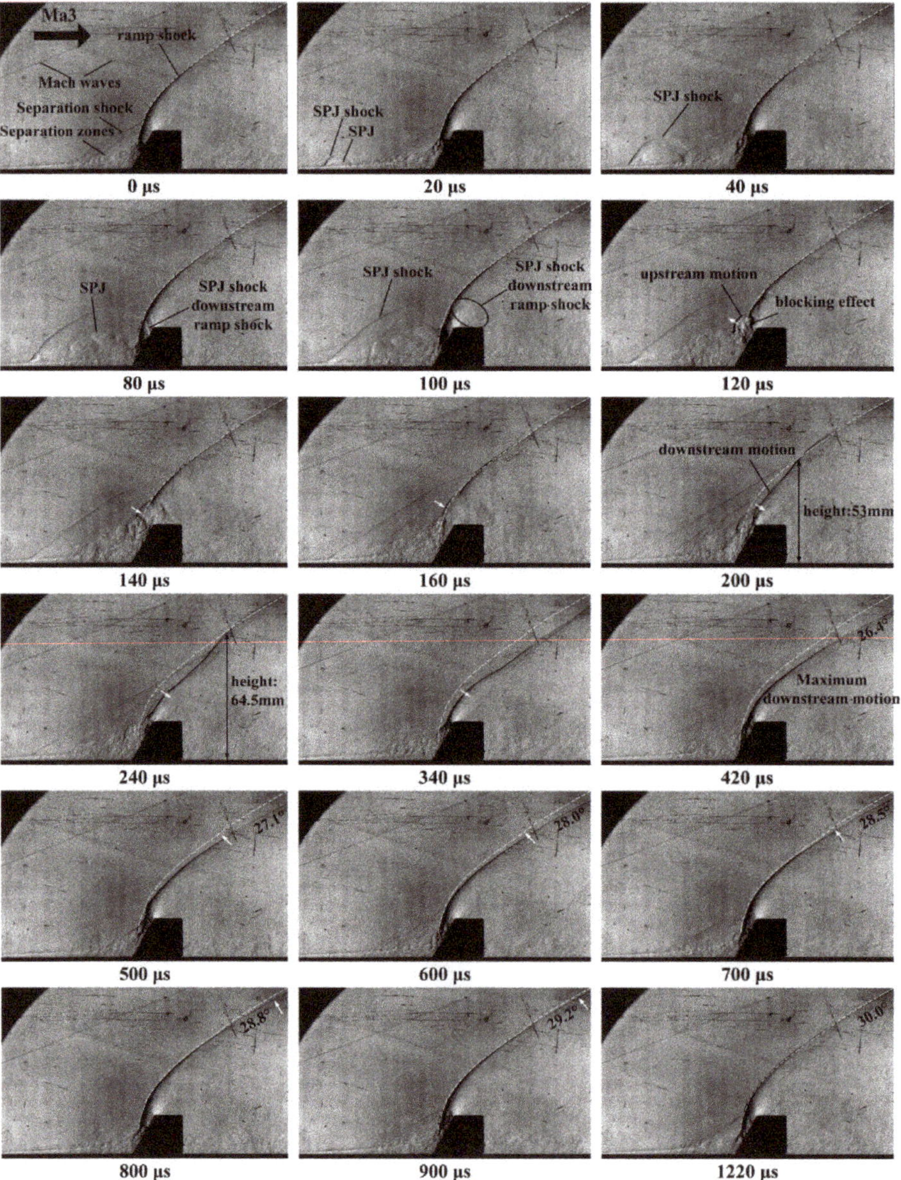

Figure 3. Interaction process between SPJ and the supersonic crossflow near the ramp in case1.

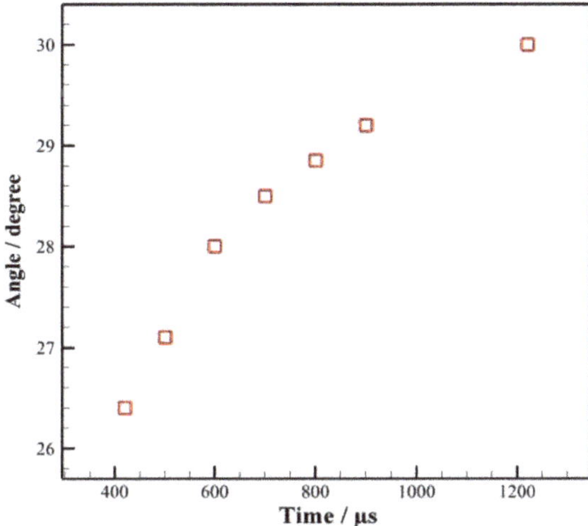

Figure 4. Change curve of the ramp shock tail angle with time during the recovery process.

Schlieren imaging is the main measurement method in the fields of shock wave control using SPJ. However, the disadvantage is obvious that only a few qualitative analysis data can be obtained. Therefore, high-frequency dynamic pressure sensor is adopted in this paper to try to measure the pressure on the ramp wall after the ramp shock, so as to prove that the ramp shock is weakened by SPJ and analyze its degree of weakening.

Change curves of discharge voltage and ramp wall pressure with time measured in case1 are shown in Figure 5. In the base flow field, the pressure of the ramp wall (the position of the pressure measuring hole) fluctuates around 30 kPa (the base pressure P_b in Figure 5), but the fluctuation range is relatively small. The voltage at both ends of the discharge capacitor is approximately zero before charging begins. About 350 μs before the start of the discharge (at t = −350 μs), the discharge capacitor begins to charge, and the voltage at both ends of the capacitor increases continuously. At 0 μs, the voltage at both ends of the capacitor reaches the breakdown voltage (about 1.91 kV), and the discharge begins. Therefore, the energy input into the discharge cavity is approximately 1.17 J. Since the time scale of discharge is much smaller than the time scale of pressure change, it is difficult to identify the voltage waveform under the time scale of the x-coordinate in Figure 5. Actually, the voltage waveform oscillates up and down and decays with time. After a period of response time ($\Delta t_d \approx$75 μs), the ramp wall pressure began to be greatly disturbed. The ramp wall pressure first rises for a short time, as shown in Figure 5, and the pressure at the peak is about 36 kPa. The appearance of the small pressure peak at 75 μs may be caused by the impacting of SPJ shock. Then, the ramp wall pressure begins to drop. After about 370 μs ($\Delta t_p \approx$ 370 μs), the ramp wall pressure reaches its minimum at about 6.3 kPa. After that, the ramp wall pressure starts to recover. Compared with the pressure drop process, the speed of the pressure recovery process is slightly slower. It takes a total of about 880 μs from initial response time t_1 to pressure recovery time t_2, that is, $\Delta t_a \approx$ 880 μs. Dynamic pressure measurement results show that, although there will be a brief pressure rise, on the whole, the ramp pressure decreases obviously under control of SPJ and SPJ shock. The minimum pressure is reduced by a maximum of 79% compared to the base pressure, which indicates that the strength of the ramp shock is significantly weakened.

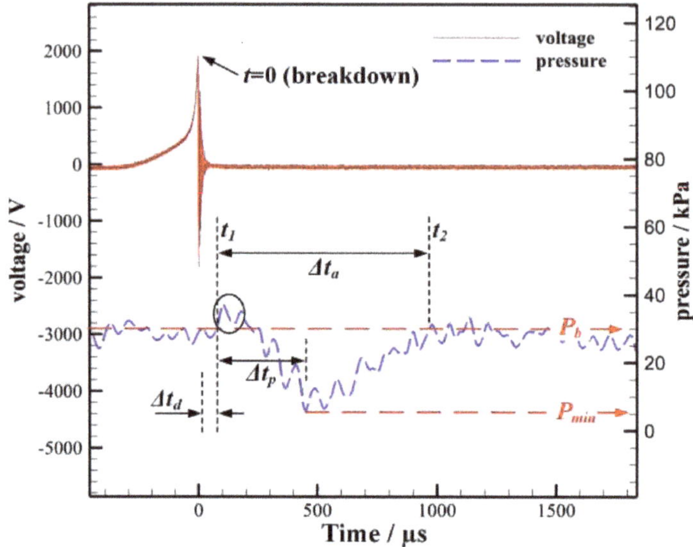

Figure 5. Change curves of discharge voltage and ramp wall pressure with time in case1.

3.2. *Effect of Discharge Capacitance*

Comparison of flow field evolution process with different discharge capacitance (case1, case2 and case3) is shown in Figure 6. The larger the discharge capacitance is, the more energy is injected into the discharge cavity and the higher the pressure rise in the discharge cavity is. Therefore, as we can see from Figure 6, the larger the discharge capacitance is, the stronger the intensity of SPJ and SPJ shock is. At 40 µs after discharge, the angle of SPJ shock in case1, case2 and case3 are 56.1, 46.3° and 36.9°, respectively. In addition, the larger the discharge capacitance is, the higher temperature and lower density SPJ will be formed, resulting in a higher density gradient and clearer display in the schlieren images. For example, at 80 µs, large scale vortex structure formed by SPJ can be clearly observed in case1, but it is difficult to be observed in case2 and case3. Velocity of SPJ and SPJ shock both in the flow direction and the normal direction increase with discharge capacitance. At 420, 500 and 520 µs, respectively, in case1, case2 and case3, the angles of ramp shock tail reach the minimum value at 26.4, 27.3 and 28.1°. From the measurement of ramp pressure, as shown in Figure 7, it can be concluded that the increase in discharge capacitance can increase the degree of ramp pressure reduction, that is, enhance the control effect of SPJ on ramp shock. The minimum ramp wall pressure values are 6.3, 12.8 and 18.7 kPa in case1, case2 and case3, respectively, which are 79.0, 57.3 and 37.7% lower than the base pressure.

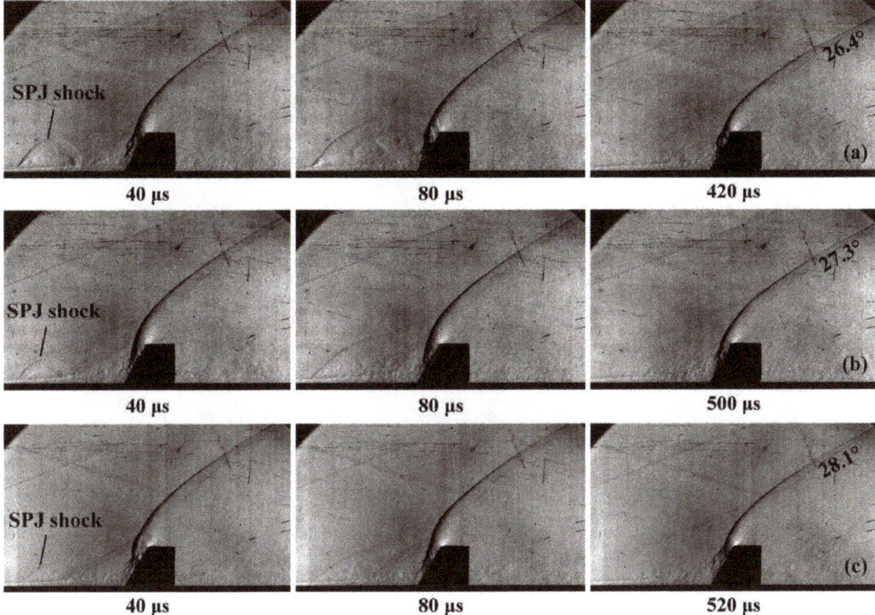

Figure 6. Comparison of the flow field evolution process with different discharge capacitance: (**a**) case1—640 nF, (**b**) case2—320 nF and (**c**) case3—80 nF.

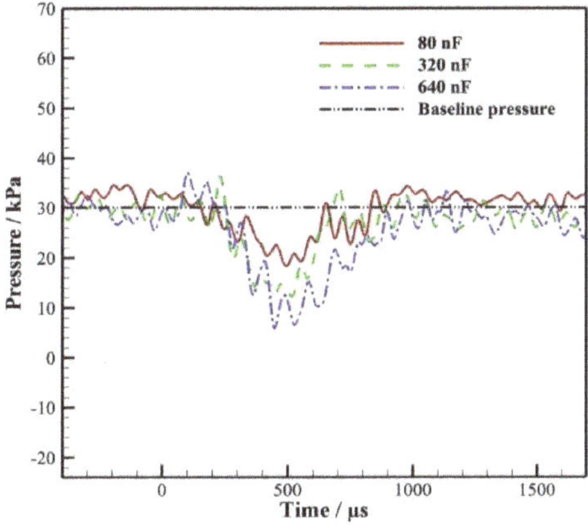

Figure 7. Change curves of ramp pressure with different discharge capacitances.

3.3. *Effect of Exit Diameter*

Comparison of flow field evolution process with different exit diameter (case4—1.5 mm, case5—5 mm and case6—11 mm) is shown in Figure 8. By comparing the flow field at 40 and 80 μs, it can be seen that the larger the exit diameter of SPJ actuator is, the stronger the SPJ shock generated is, and the faster the SPJ shock moves. For example, at 40 μs, the SPJ shock angle in case4, case5 and case6 is 45, 49 and 52.5°, respectively. For case6, in addition to the relatively strong SPJ shock, a weak separation oblique shock is generated upstream the SPJ exit. The angle of the oblique shock is about 22°, which is close to (slightly greater than) the interference shock angle (21.5°) generated by the uneven wall surface in Ma3 flow. While in case5, the separation oblique shock is very weak, and in case4, it cannot be observed.

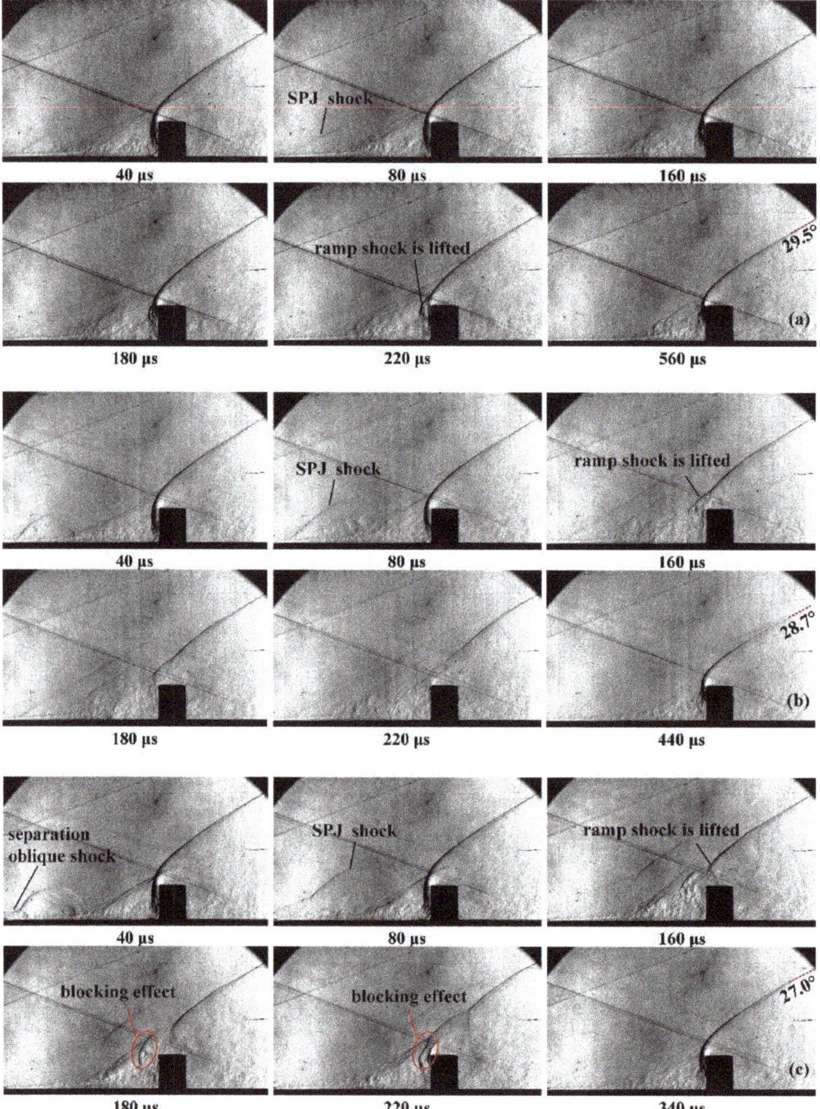

Figure 8. Comparison of the flow field evolution process with different exit diameters: (**a**) case4—1.5 mm, (**b**) case5—5 mm, (**c**) case6—11 mm.

As shown in Figure 9, the minimum ramp wall pressures in case4, case5 and case6 are 12.6, 6.4 and 9.7 kPa, respectively, which are 59.7, 79.5 and 69% lower than the base pressure (31.3 kPa for the 90° ramp). The results show that when the exit diameter increases from 5 to 11 mm, the effect of shock weakening will decrease. Analysis from schlieren images shows that that the 11 mm diameter actuator produces SPJ with higher mass flow, which has a relatively strong blocking effect on the flow field. As shown in Figure 8c, at 180 and 220 μs, a strong shock is generated in front of the ramp due to the blocking effect, which is just in front of the pressure measurement hole. Therefore, to a certain extent, the effect of the shock weakening is reduced.

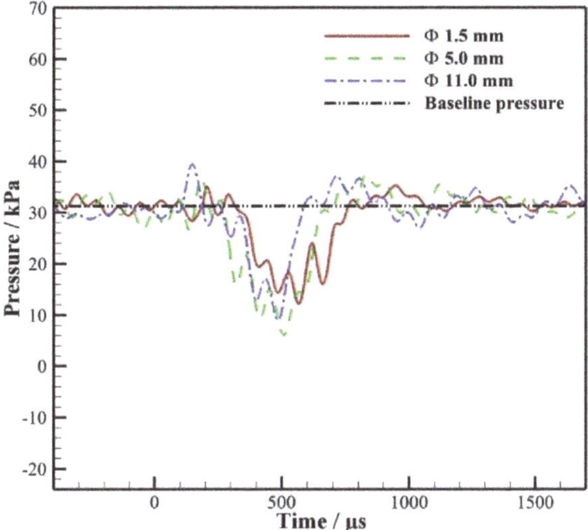

Figure 9. Change curves of ramp pressure with different exit diameter.

For the time scale, the larger the diameter of SPJ exit is, the earlier the control effect and the maximum effect are generated, but also the earlier they end. Therefore, it can be considered that the "phase" of the control effect is advanced along with the increase in exit diameter. It can be seen from change curves in Figure 9 that the larger the exit diameter is, the earlier the moment of the "pressure peak", the moment of the pressure minimum and the moment of the pressure recovery appear. From schlieren images in Figure 8, we can see that for case5 and case6, the ramp shock is lifted and partly eliminated at 160 μs. However, for case4, it is not until 180 μs when the ramp shock begins to show minor changes, and 220 μs when the ramp shock is lifted and partly eliminated. A similar rule appears in the moment of minimum ramp shock tail angle, of which the moments are, respectively, 560, 440 and 340 μs in case4, case5 and case6.

3.4. *Effect of Ramp Distance*

Characteristics of the SPJ actuator and the changes in the ramp shock are significantly different when the SPJ exit is located in the separation zone and outside. Therefore, a comparison is made between the two situations in case7 and case5 first. The electromagnetic interference due to the discharge is too strong to measure the change in ramp wall pressure when SPJ actuator is in the separation zone in case7. Therefore, comparison is done through schlieren images in Figure 10. The ramp angle in the two cases is 90°.

When the SPJ exit is located in the separation zone in case7, the initial pressure in the actuator cavity is higher than that outside the separation zone in case5 due to the decrease in flow velocity and the increase in flow pressure. As a result, the breakdown voltage increases from about 1.9 kV in case5 to about 3.4 kV in case7. As shown in Figure 10a, at 20 μs, the discharge arc is ejected out of SPJ exit due to the low velocity and rotational flow in the separation zone. The time scale is also significantly affected by the separation zone. For case7, the ramp shock is significantly changed at 40 μs, and the maximum change in the ramp shock tail angle appears at 200 μs. However, for case5, the SPJ and SPJ shock are still in the development stage before 140 μs and the maximum change in the ramp shock tail angle appears at 440 μs.

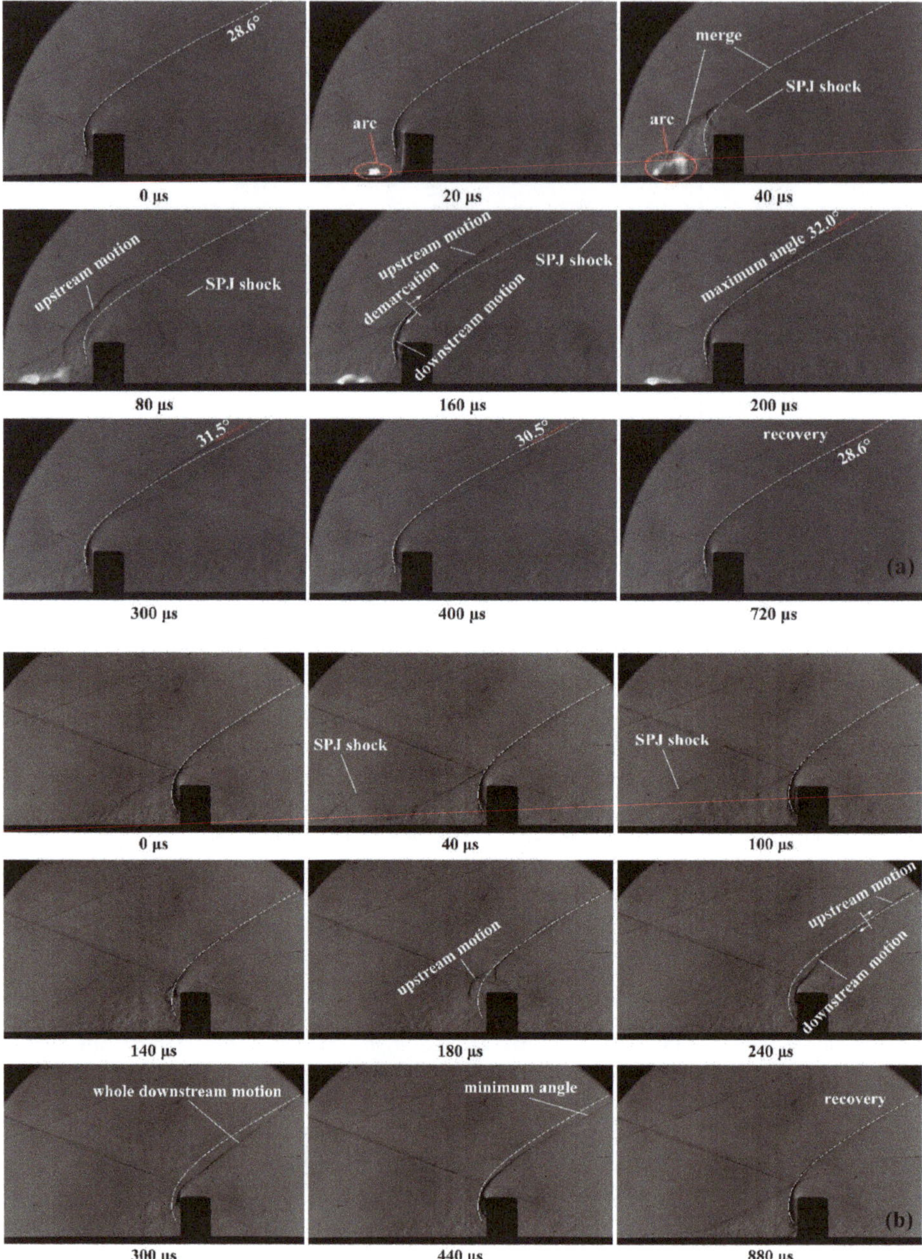

Figure 10. Comparison of the flow field evolution process when SPJ exit is in the separation zone and outside: (**a**) case7—in the separation zone, (**b**) case5—outside the separation zone.

In the initial stage, the control effect in the separation zone is similar to that outside the separation zone. As shown in Figure 10a at 40 μs and Figure 10b at 180 μs, the ramp shock root is lifted and partly eliminated under the control of the high-temperature SPJ. However, there are also some differences.

The SPJ shock in case7 is directly fused with the ramp shock and the pressure disturbance of SPJ shock on ramp shock is applied from the downstream of ramp shock, which spreads along the flow direction and normal direction (equally important). The effect is similar to that of an opposing jet. Instead, when SPJ exit is outside the separation zone in case5, the pressure disturbance of the SPJ shock on the ramp shock is applied from the upstream of ramp shock, and the disturbance gradually spreads along the flow direction and normal direction, mainly in the flow direction.

Therefore, affected by the initial stage, effects of SPJ and SPJ shock on the ramp shock in the later stage are very different. In case7, as shown in Figure 10a, at 80, 160 and 200 μs, SPJ shock and the ramp shock merge and are lifted. As the disturbance expands in the normal direction, a lifted part of the ramp shock gradually expands, at the same time, the angle of the shock tail gradually increases. At 200 μs, the ramp shock in the observed area is completely lifted, and the angle of the shock tail reaches its maximum value of 32°. After that, the ramp shock gradually recovers, and the angle of the ramp shock tail gradually decreases. At 720 μs, the ramp shock basically returns to the base flow field position. In case5, similar to case1 described in Section 3.1, the ramp shock first goes through a process of "short-term local upstream motion", as shown in Figure 10b at 180 μs. After that, the ramp shock is mainly represented by "long-term whole downstream motion", as shown in Figure 10b at 240, 300 and 440 μs. At 880 μs, the ramp shock basically returns to the base flow field position.

When the SPJ exit is outside the separation zone, with the change in the ramp distance, the control effect is similar, but the degree is different. Figure 11 shows the comparison of the flow field evolution process under the control of SPJ of SPJ shock when the SPJ exit is located outside the separation zone at different ramp distances. Ramp angles in these cases are 60°. Ramp distances in case8, case1, case9 and case10 are, respectively, 30, 50, 70, 90 mm.

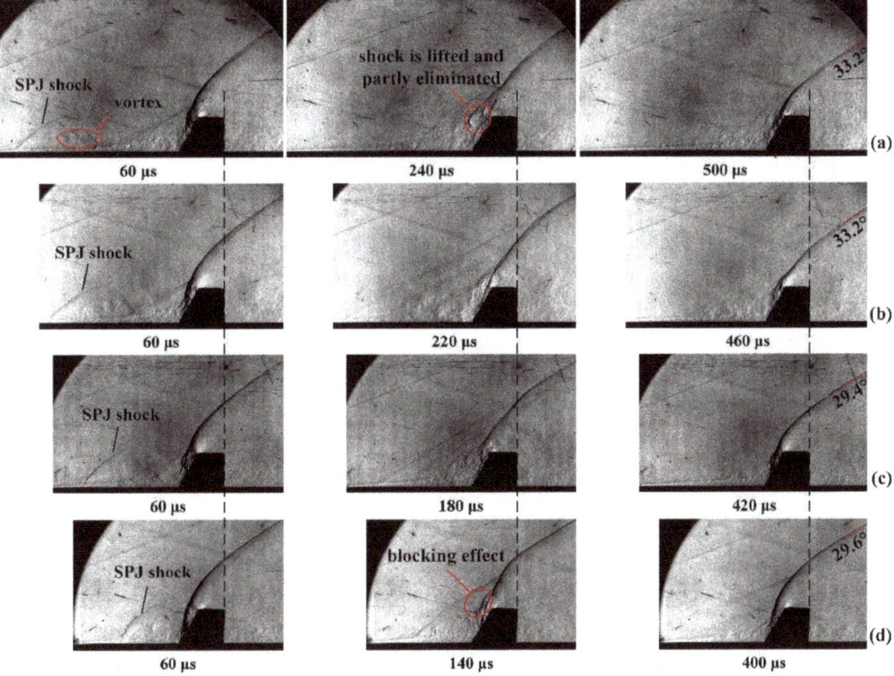

Figure 11. Comparison of the flow field evolution process when SPJ exit is outside the separation zone at different ramp distances: (**a**) case10—90 mm, (**b**) case9—70 mm, (**c**) case1—50 mm, (**d**) case8—30 mm.

The SPJ shock generated in case9 and case10 is similar. However, in case1 and case8, due to that it is closer to the separation zone with higher static pressure, the breakdown voltage and the discharge energy increases, SPJ shock is strengthened along with the decrease in ramp distance. Then, change curves of ramp pressure in the four cases are compared in Figure 12. The minimum ramp wall pressures are 14.1, 6.3, 13.1 and 14.6 kPa in case8, case1, case9, case10, respectively, which are reduced by 53.0, 79.0, 56.3 and 51.3% compared to the base pressure, respectively. When the ramp distance increases from 50 to 90 mm, intensity of SPJ and SPJ shock weakens during the long-distance movement, which will weaken the control effect on the ramp shock. However, when the ramp distance decreases from 50 to 30 mm, control effect is also weakened. The reason may be similar to that in case6, namely a new strong shock is produced upstream the ramp, as shown in Figure 11d, at 140 μs, which increases the ramp wall pressure.

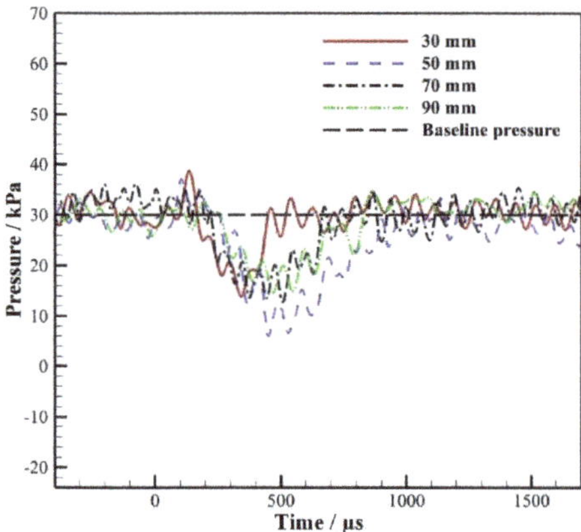

Figure 12. Change curves of ramp pressure with different ramp distances.

From the perspective of the time scale, the smaller the ramp distance is, the faster the control effect appears. The lifting effect of the ramp shocks in case8, case1, case9 and case10 reaches the maximum at 140, 180, 220 and 240 μs, respectively, and the reduction in the angle of the ramp shock tail reaches the minimum at 400, 420, 460 and 500 μs, respectively, as shown in Figure 11. However, change curves of ramp pressure in Figure 12 show that, in terms of the duration of control effect, it is still the longest when the ramp distance is 50 mm. The analysis suggests that this may be because the control effect is weak when the ramp distance is relatively longer, so it attenuates quickly.

4. Conclusions

In this paper, control of ramp shock wave in Ma3 supersonic flow using two-electrode SparkJet (SPJ) actuator is investigated experimentally by using schlieren images and dynamic pressure measurement results. The main conclusions are as follows:

1. Under control of SPJ and SPJ shock, not only the angle and position of the ramp shock are changed, but also the intensity is weakened. The measurement results of the ramp wall pressure show that the ramp pressure is reduced by a maximum of 79% compared to the pressure in the base flow field.
2. Ten experimental cases are set for investigating and analyzing the effects of some parameters including discharge capacitance, exit diameter and ramp distance on the control effect of SPJ on the ramp shock in detail. The increase in discharge capacitance helps to improve the control effect of SPJ on the ramp shock. However, the control effect of SPJ actuator with medium exit diameter is better than that with too small or too large one. In addition, when the SPJ exit is located in the

separation zone and outside, the change in the ramp shock shows significant differences, but the control effect on the ramp shock in the case of medium ramp distance is better when the SPJ exit is located outside the separation zone.

Author Contributions: Conceptualization, W.X. and Y.Z.; methodology, Z.L.; validation, L.W. and W.P.; writing—original draft preparation, W.X.; writing—review and editing, Y.Z. and T.G. All authors have read and agreed to the published version of the manuscript.

Funding: The present study was funded by the National Natural Science Foundation of China (Grant No. 12002377, 11972369, 11872374, and 52075538), the Natural Science Foundation of Hunan Province (Grant No. 2020JJ5670) and research program of National University of Defense Technology (Grant No. ZK18-03-11).

Conflicts of Interest: The authors declare no conflict of interests.

References

1. Meng, X.S.; Hu, H.Y.; Li, C.; Abbasi, A.A.; Cai, J.S.; Hu, H. Mechanism study of coupled aerodynamic and thermal effects using plasma actuation for anti-icing. *Phys. Fluids* **2019**, *31*, 037103. [CrossRef]
2. Gao, T.X.; Luo, Z.B.; Zhou, Y.; Liu, Z.Y.; Peng, W.Q.; Cheng, P.; Deng, X. Novel deicing method based on plasma synthetic jet actuator. *AIAA J.* **2020**, *58*, 1–8. [CrossRef]
3. Gao, T.X.; Luo, Z.B.; Zhou, Y.; Yang, S.K. A novel de-icing strategy combining electric-heating with plasma synthetic jet actuator. *Proc. Inst. Mech. Eng. Part G J. Aerosp. Eng.* **2020**. [CrossRef]
4. Chen, W.Q.; Jin, D.; Cui, W.; Huang, S.F. Characteristics of gliding arc plasma and its application in swirl flame static instability control. *Processes* **2020**, *8*, 684. [CrossRef]
5. Liu, C.Y.; Sun, M.B.; Wang, H.B.; Yang, L.C.; An, B.; Pan, Y. Ignition and flame stabilization characteristics in an ethylene-fueled scramjet combustor. *Aerosp. Sci. Technol.* **2020**, *106*, 106186. [CrossRef]
6. Wang, J.J.; Choi, K.S.; Feng, L.H.; Jukes, T.N.; Whalley, R.D. Recent developments in DBD plasma flow control. *Prog. Aerosp. Sci.* **2013**, *62*, 52–78. [CrossRef]
7. Tang, M.X.; Wu, Y.; Guo, S.G.; Sun, Z.Z.; Luo, Z.B. Effect of the streamwise pulsed arc discharge array on shock wave/boundary layer interaction control. *Phys. Fluids* **2020**, *32*, 076104. [CrossRef]
8. Popkin, S.H.; Cybyk, B.Z.; Land, B.; Foster, C.H.; Alvi, F.S. Recent performance-based advances in SparkJet actuator design for supersonic flow applications. In Proceedings of the 51st AIAA Aerospace Sciences Meeting Including the New Horizons Forum and Aerospace Exposition, Grapevine, TX, USA, 7–10 January 2013.
9. Zong, H.H.; Kotsonis, M. Formation, evolution and scaling of plasma synthetic jets. *J. Fluid Mech.* **2018**, *837*, 147–181. [CrossRef]
10. Grossman, K.R.; Cybyk, B.Z.; VanWie, D.M. SparkJet actuators for flow control. In Proceedings of the 41st Aerospace Sciences Meeting and Exhibit, Reno, NV, USA, 6–9 January 2003.
11. Reedy, T.M.; Kale, N.V.; Dutton, J.C.; Elliott, G.S. Experimental characterization of a pulsed plasma jet. *AIAA J.* **2013**, *51*, 2027–2031. [CrossRef]
12. Wang, L.; Xia, Z.X.; Luo, Z.B.; Zhou, Y.; Zhang, Y. Experimental study on the characteristics of a two-electrode plasma synthetic jet actuator. *Acta Phys. Sin.* **2014**, *63*, 194702.
13. Huang, H.X.; Tan, H.J.; Guo, Y.J.; Sun, S.; He, X.M. Flowfield induced by a plasma synthetic jet actuator with low exit inclination angle under low ambient pressure. *Aerosp. Sci. Technol.* **2020**, *105*, 106018. [CrossRef]
14. Li, J.; Zhang, X. A novel model of plasma synthetic jet actuators coupled energy source term model with mechano-acoustical analogy model. *J. Phys. D Appl. Phys.* **2020**, *53*, 235204. [CrossRef]
15. Seyhan, M.; Akansu, Y.E. The effect of a novel spark-plug plasma synthetic jet actuator on the performance of a PEM fuel cell. *Int. J. Heat Mass Transf.* **2019**, *140*, 147–151. [CrossRef]
16. Chedevergne, F.; Bodoc, V.; Leon, O.; Caruana, D. Experimental and numerical response of a high-Reynolds-number M = 0.6 jet to a plasma synthetic jet actuator. *Int. J. Heat Fluid Flow* **2015**, *56*, 1–15. [CrossRef]
17. Dufour, G.; Hardy, P.; Quint, G.; Rogier, F. Physics and models for plasma synthetic jets. *Int. J. Aerodyn.* **2013**, *3*, 47–70. [CrossRef]
18. Takehiro, H.; Kakuji, O.; Shinsuke, M. Direct Measurement of Wall-Shear Stress of Plane Shear Layer with Plasma Synthetic Jet Actuator. *J. Fluid Sci. Technol.* **2009**, *4*, 75–83.

19. Ricchiuto, A.C.; Borghi, C.A.; Cristofolini, A.G.; Neretti, G. Measurement of the charge distribution deposited on a target surface by an annular plasma synthetic jet actuator: Influence of humidity and electric field. *J. Electrost.* **2020**, *107*, 103501. [CrossRef]
20. Zhang, P.F.; Dai, C.F.; Liu, A.B.; Wang, J.J. The effect of actuation frequency on the plasma synthetic jet. *Sci. China Tech. Sci.* **2011**, *54*, 2945. [CrossRef]
21. Wu, S.; Liu, X.; Huang, G.; Liu, C.; Bian, W.; Zhang, C. Influence of high-voltage pulse parameters on the propagation of a plasma synthetic jet. *Plasma Sci. Technol.* **2019**, *21*, 074007. [CrossRef]
22. Zong, H.H.; Kotsonis, M. Effect of velocity ratio on the interaction between plasma synthetic jets and turbulent cross-flow. *J. Fluid Mech.* **2019**, *865*, 928–962. [CrossRef]
23. Zhang, Y.C.; Tan, H.J.; Huang, H.X.; Sun, S.; He, X.M.; Cheng, L.; Zhuang, Y. Transient flow patterns of multiple plasma synthetic jets under different ambient pressures. *Flow Turbul. Combust.* **2018**, *101*, 741–757. [CrossRef]
24. Zhang, W.; Geng, X.; Shi, Z.W.; Jin, S.L. Study on inner characteristics of plasma synthetic jet actuator and geometric effects. *Aerosp. Sci. Technol.* **2020**, *105*, 106044. [CrossRef]
25. Yu, Y.; Xu, J.L.; Gan, N. Effects of Parameters on Continuously Working Plasma Synthetic Jet. *Eng. Appl. Comput. Fluid Mech.* **2014**, *8*, 55–69. [CrossRef]
26. Sary, G.; Dufour, G.; Rogier, F.; Kourtzanidis, K. Modeling and parametric study of a plasma synthetic jet for flow control. *AIAA J.* **2014**, *52*, 1591–1603. [CrossRef]
27. Haack, S.J.; Taylor, T.M.; Cybyk, B.Z.; Foster, C.H.; Alvi, F.S. Experimental estimation of SparkJet efficiency. In Proceedings of the 42nd AIAA Plasmadynamics and Lasers Conference in Conjunction with the 18th International Conference on MHD Energy Conversion (ICMHD), Honolulu, HI, USA, 27–30 June 2011.
28. Golbabaei-Asl, M.; Knight, D.; Wilkinson, S. Novel technique to determine SparkJet efficiency. *AIAA J.* **2015**, *53*, 501–882. [CrossRef]
29. Wang, L.; Xia, Z.X.; Luo, Z.B.; Chen, J. Three-electrode plasma synthetic jet actuator for high-speed flow control. *AIAA J.* **2014**, *52*, 879–882. [CrossRef]
30. Neretti, G.; Seri, P.; Taglioli, M.; Shaw, A.; Iza, F.; Borghi, C.A. Geometry optimization of linear and annular plasma synthetic jet actuators. *J. Phys. D Appl. Phys.* **2017**, *50*, 015210. [CrossRef]
31. Caruana, D.; Barricau, P.; Hardy, P. The "plasma synthetic jet" actuator aero-thermodynamic characterization and first flow control applications. In Proceedings of the 47th AIAA Aerospace Sciences Meeting Including the New Horizons Forum and Aerospace Exposition, Orlando, FL, USA, 5–8 January 2009.
32. Liu, R.B.; Niu, Z.G.; Wang, M.M.; Hao, M.; Lin, Q. Aerodynamic control of NACA 0021 airfoil model with spark discharge plasma synthetic jets. *Sci. China Tech. Sci.* **2015**, *58*, 1949–1955. [CrossRef]
33. Anderson, K.V.; Knight, D.D. Plasma jet for flight control. *AIAA J.* **2012**, *50*, 1855–1872. [CrossRef]
34. Wang, P.; Shen, C.B. Characteristics of mixing enhancement achieved using a pulsed plasma synthetic jet in a supersonic flow. *J. Zhejiang Univ. Sci. A* **2019**, *20*, 701–713. [CrossRef]
35. Caruana, D.; Barricau, P.; Gleyzes, C. Separation control with plasma synthetic jet actuators. *Int. J. Aerodyn.* **2013**, *3*, 71–83. [CrossRef]
36. Wang, H.Y.; Li, J.; Jin, D.; Tang, M.X.; Wu, Y.; Xiao, L. High-frequency counter-flow plasma synthetic jet actuator and its application in suppression of supersonic flow separation. *Acta Astronaut.* **2017**, *142*, 45–56. [CrossRef]
37. Ogawara, K.; Kojima, R.; Matsumoto, S.J.; Shingin, H. Extremum seeking adaptive separation control on a wing with plasma synthetic jet actuator. *J. Fluid Sci. Technol.* **2012**, *7*, 89–99. [CrossRef]
38. Zong, H.H.; Van, P.T.; Kotsonis, M. Airfoil flow separation control with plasma synthetic jets at moderate Reynolds number. *Exp. Fluids* **2018**, *59*, 169. [CrossRef]
39. Cybyk, B.Z.; Wilkerson, J.; Grossman, K.R. Performance characteristics of the SparkJet flow control actuator. In Proceedings of the 2nd AIAA Flow Control Conference, Portland, OR, USA, 28 June–1 July 2004.
40. Narayanaswamy, V.; Raja, L.L.; Clemens, N.T. Characterization of a high-frequency pulsed-plasma jet actuator for supersonic flow control. *AIAA J.* **2010**, *48*, 297–305. [CrossRef]
41. Narayanaswamy, V.; Raja, L.L.; Clemens, N.T. Control of unsteadiness of a shock wave/turbulent boundary layer interaction by using a pulsed-plasma-jet actuator. *Phys. Fluids* **2012**, *24*, 543. [CrossRef]
42. Narayanaswamy, V.; Raja, L.L.; Clemens, N.T. Control of a Shock/Boundary-Layer Interaction by using a pulsed-plasma jet actuator. *AIAA J.* **2012**, *50*, 246–249. [CrossRef]

43. Wang, H.Y.; Li, J.; Jin, D.; Dai, H.; Gan, T.; Wu, Y. Effect of a transverse plasma jet on a shock wave induced by a ramp. *Chin. J. Aeronaut.* **2017**, *30*, 1854–1865. [CrossRef]
44. Huang, H.X.; Tan, H.J.; Sun, S.; Zhang, Y.C.; Cheng, L. Transient interaction between plasma jet and supersonic compression ramp flow. *Phys. Fluids* **2018**, *30*, 041703. [CrossRef]
45. Zhou, Y.; Xia, Z.X.; Luo, Z.B.; Wang, L. Effect of three-electrode plasma synthetic jet actuator on shock wave control. *Sci. China Technol. Sci.* **2017**, *60*, 150–156. [CrossRef]
46. Zhou, Y.; Xia, Z.X.; Luo, Z.B.; Wang, L.; Deng, X.; Zhang, Q.H.; Yang, S.K. Characterization of three-electrode SparkJet actuator for hypersonic flow control. *AIAA J.* **2019**, *57*, 879–885. [CrossRef]
47. Wang, D.; Zhao, Y.; Xia, Z.; Wang, Q.; Huang, X. Multi-resolution analysis of density fluctuation of coherent structures about supersonic flow over vg. *Chin. J. Aeronaut.* **2012**, *2*, 173–181. [CrossRef]

Article

Characteristics of Gliding Arc Plasma and Its Application in Swirl Flame Static Instability Control

Weiqi Chen [1], Di Jin [1,2,*], Wei Cui [1,3] and Shengfang Huang [1]

1 Science and Technology on Plasma Dynamics Laboratory, Air Force Engineering University, Xi'an 710038, China; chenweiqi@mail.nwpu.edu.cn (W.C.); cuiwei@xjtu.edu.cn (W.C.); shengfang_huang@126.com (S.H.)
2 Department of Energy and Power Engineering, Tsinghua University, Beijing 100084, China
3 Institute of Aero-Engine, School of Mechanical Engineering, Xi'an Jiaotong University, Xi'an 710049, China
* Correspondence: james.jd@163.com; Tel.: +86-1399-111-7151

Received: 3 May 2020; Accepted: 5 June 2020; Published: 11 June 2020

Abstract: Based on an experimental system involving a pulsating airflow burner and gliding arc generator, the characteristics of gliding arc plasma at different flow rates and its control effect on the static instability of the swirl flame have been studied. The current, voltage, and power wave forms, as well as the simultaneous evolution of plasma topology, were measured to reveal the discharge characteristics of the gliding arc. A bandpass filter was used to capture the chemiluminescence of CH in the flame, and pressure at the burner outlet was acquired to investigate the static instability. Experimental results showed that there were two different discharge types in gliding arc plasma. With the low flow rate, the glow type discharge was sustained and the current was nearly a sine wave with hundreds of milliamperes of amplitude. With the high flow rate, the spark type discharge appeared and spikes which approached almost 1 ampere in 1 μs were found in the current waveform. The lean blowout limits increased when the flame mode changed from stable to pulsating, and decreased significantly after applying the gliding arc plasma. In pulsating flow mode, the measured pressure indicated that static instability was generated at the frequency of 10 Hz, and the images of flame with plasma showed that the plasma may have acted as the ignition source which injected the heat into the flame.

Keywords: gliding arc plasma; discharge characteristics; swirl flame; static instability control

1. Introduction

With the development of the aviation industry and the improvement of environmental awareness, exhaust emission standards for NOx, CO, and other contaminants are becoming stricter. Thus, lean premixed combustion technology has been widely adopted in aircraft engine combustion chambers to reduce NOx [1]. However, the instability of premixed flame is severer than that of non-premixed flame. When the aircraft is maneuvering in flight, the air inflow of the combustor may suddenly fluctuate, and the transient equivalence ratio may sharply fall below the lean blowout limit, resulting in flame extinguishment and a threat to flight safety. Consequently, it is crucial to control the instability of the lean premixed flame to improve the reliability of aero-engine combustion chambers.

Adopting swirl flow is the general method for improving flame stability [2]. The swirl flow can generate a recirculation zone which makes the hot burned gas flow back to the bottom of the flame and ignite the fresh air. According to previous research, flame instability can be classified into dynamic instability and static instability [3]. Dynamic instability usually refers to combustion-driven oscillations which involve thermoacoustic coupling and fuel/air wave coupling, while static instability is considered to be related to flame blowout [4]. For example, when the aircraft takes off or lands, the air flowing through the engine will probably be fluctuant, and the flame in the combustor may be extinguished transiently. Then, the thermal energy generated by the combustor will be reduced, causing a decline in the rotational speeds of the turbine and compressor, and the air absorbed by compressor will be reduced accordingly. The reduced air inflow will increase the equivalence ratio and the flame will be reignited. Thus, the flame will go into a cycle of extinguishment and re-ignition, and static instability will be generated. In 1878, Rayleigh proposed a theoretical model of oscillation combustion relating to the interaction of acoustic instability and thermodynamics, which was later summarized as the Rayleigh Criterion. According to the criterion, combustion oscillation usually occurs when the phase difference between the pressure wave and heat release is less than 90° [5]. Based on this research result, more researchers focused on predicting combustion oscillation and developed flame transfer function methods to calculate the resonant frequencies of flame oscillation [6]. Controlling flame instability and then reducing the lean blowout limits was another issue that researchers focused on. The geometry of combustor was the primary parameter that was optimized to achieve the best extinction performance [7]. Then, other passive methods to control the thermoacoustic instability of flame were summarized [8]. Recently, the plasma control method has attracted the attention of many researchers. As a kind of high-temperature and reactive material, plasma has played an important role in assisting combustion [9]. Some researchers applied plasma to supersonic combustion [10,11], turbulent premixed flames [12], and counterflow diffusion flames [13–15], and found that plasma had a good effect in ignition and controlling instability. Moreover, many types of discharge methods, such as gliding arc discharges [16,17], nanosecond pulsed discharges [18], microwave discharges [19], and radio frequency discharges [20], proved helpful to combustion.

As non-equilibrium plasma, the gliding arc can produce reactive species in atmospheric pressure easily [21,22] and can be widely used in assisting combustion. The main species generated by the gliding arc contain NO*, N_2*, and OH*, and their spatial distributions are quite different [23]. The characteristics of gliding arc plasma and its applications have been investigated by many researchers, but there are few reports applying it in flame instability control.

In this paper, an experimental system was designed whereby the swirl flame was excited by pulsating airflow, and the gliding arc plasma was arranged at the burner outlet so the gas could blow gliding arc out from the electrodes and the swirl flame could interact directly with the gliding arc plasma. The characteristics of the gliding arc, as well as its control effect on the static instability of premixed swirl flame, were investigated experimentally in the research, and an extension of lean blowout limits for the methane–air flame was observed. Specifically, the evolution process of plasma topology in one discharge period was captured firstly, and the voltage, current, and power waveforms were recorded simultaneously, which provided a theoretical basis for controlling the flame static instability by plasma. Then, the lean blowout limits of swirl flame in stable flow mode, pulsating flow mode, and pulsating flow mode with gliding arc plasma were measured, leading to the conclusion that the extinction performance of pulsating flame coupled with plasma was better than that of the pulsating flame or stable flame. Finally, the relationships between pressure and heat release rate of the pulsating flame were measured to explore the mechanism of the static instability.

2. Experimental Facility

Figure 1 shows a schematic of experimental system, including the pulsating air flow generator, multi-swirl burner, gliding arc plasma power supply, and other measuring devices.

2.1. Pulsating Air Flow Generator

A schematic diagram of the pulsating air flow generator is shown in Figure 2.

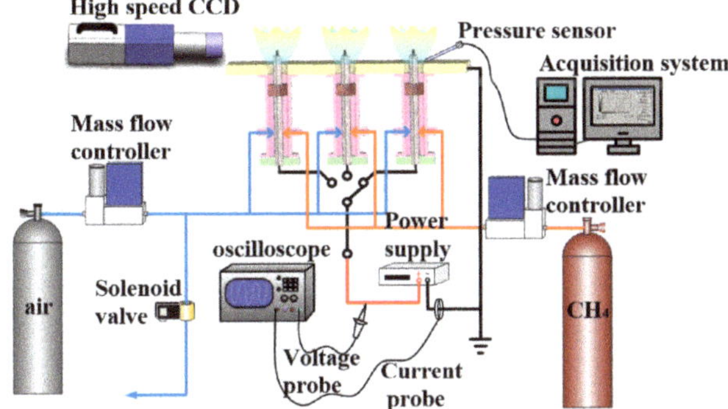

Figure 1. A schematic of the experimental system.

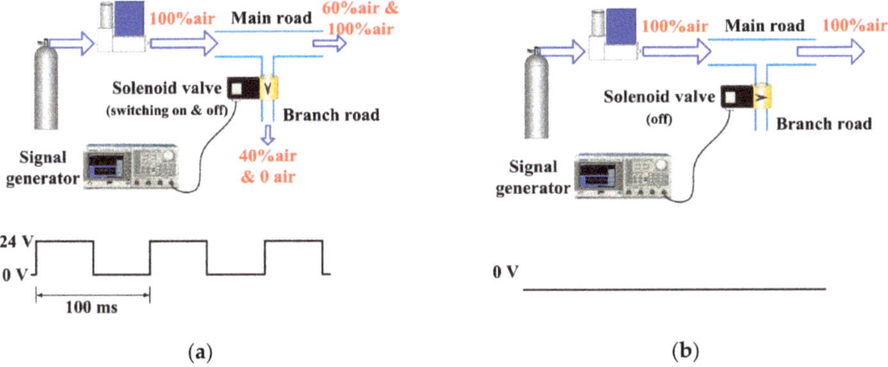

Figure 2. The working progress of pulsating air flow generator (**a**) in pulsating flow mode, and (**b**) in stable flow mode.

The air from high pressure tank firstly went into the mass flow controller (MFC), in which the air flow rate could be exactly controlled. Then it was divided into the main flow and the bypass flow. The main flow entered the multi-burner and was mixed with CH_4 to form the swirl flame. The bypass air was injected to the atmosphere without combustion. The flux of the bypass flow was controlled by a solenoid valve. When the solenoid valve was opened completely, the flux of the bypass flow was up to 40% of the total air, which meant only 60% of the air went into the burner as oxidant. When the valve was closed, 100% air could pass through the main route and the equivalence ratio (represented by variable φ), calculated as Formula (1), was lower than with the opened solenoid valve:

$$\varphi = \frac{\left(\dot{m}_{CH_4}/\dot{m}_{air}\right)}{\left(\dot{m}_{CH_4}/\dot{m}_{air}\right)_{stoic}} = \frac{\left(\dot{m}_{CH_4}/\dot{m}_{air}\right)}{0.058} = \frac{\left(\frac{\dot{V}_{CH_4}}{\rho_{CH_4}}\right)/\left(\frac{\dot{V}_{air}}{\rho_{CH_4}}\right)}{0.058} = \frac{\left(\dot{V}_{CH_4}/\dot{V}_{air}\right)_{SLM}}{0.104} \quad (1)$$

In Formula (1), $\dot{m}_{CH_4}$ represents the CH_4 mass flow rate, and $\dot{V}_{CH_4}$ represents the CH_4 volume flow rate. Similarly, the $\dot{m}_{air}$ represents the air mass flow rate and the $\dot{V}_{air}$ represents the air volume flow rate. The value of $\dot{V}_{CH_4}$ and $\dot{V}_{air}$ was recorded by MFCs. The subscript *stoic* refers to the ratio of CH_4: air in the stoichiometric condition, and the subscript *SLM* (standard liters per minute) is the unit of $\dot{V}_{CH_4}$ or $\dot{V}_{air}$. It should be noted that the φ here was calculated from the air and CH_4 bottles without consideration of the entrained air from surroundings.

The variation of φ was determined by the solenoid valve. The φ increased when the solenoid valve was closed, while it decreased when the valve was opened. The solenoid valve was controlled by a UNI-TREND UTG2025A signal generator, which output a square wave with the period of 100 ms and amplitude of 0~24 V. When the voltage reached 24 V, the solenoid valve was opened, and the bypass flow (40% of the total air) went into the atmosphere; the remaining flow (60% of the total air) was provided to the burner. When the voltage reached 0 V, the solenoid valve was closed, and 100% of the air flow went through the main route. Thus, the pulsating flow was generated to create an unstable combustion, which was defined as the pulsating flow mode (Figure 2a).

In stable flow mode the signal generator was shut down, meaning that the solenoid valve was closed permanently and the flame was kept steady with 100% of the total air, as shown in Figure 2b.

The φ constantly changed in pulsating flow mode, but in stable flow mode the φ kept steady, so it was difficult to compare the flame extinction between the two modes. In this experiment, the operating condition of pulsating air flow was regarded as being the same as that of stable air flow when the integral of the mass flow rate in pulsating flow mode with one period was equal to the integral of the flow rate in stable flow mode (Formula (2)). Then, the average air flow rate which was provided into the burner in the pulsating flow mode (represented by $(\overline{\dot{V}_{air}})_{pulsating}$) equaled that in the stable flow mode (represented by $(\overline{\dot{V}_{air}})_{stable}$), even if the instantaneous flow rates in pulsating ($(\dot{V}_{air})_{pulsating}$) and stable ($(\dot{V}_{air})_{stable}$) modes were not comparable, as shown in Figure 3. Similarly, the average of φ (represented by $\overline{\varphi}$) was calculated as per Formula (3). The response time of the solenoid valve was 2 ms, which could be ignored compared with the 100-ms pulsating period.

$$\left(\overline{\dot{V}_{air}}\right)_{stable} = \frac{\int_{t_0}^{t_0+100} \left(\dot{V}_{air}\right)_{stable} dt}{100} = \frac{\int_{t_0}^{t_0+100} \left(\dot{V}_{air}\right)_{plusating} dt}{100} = \left(\overline{\dot{V}_{air}}\right)_{plusating} \tag{2}$$

$$\overline{\varphi} = \frac{\overline{\dot{V}_{CH_4}} / \overline{\dot{V}_{air}}}{0.104} \tag{3}$$

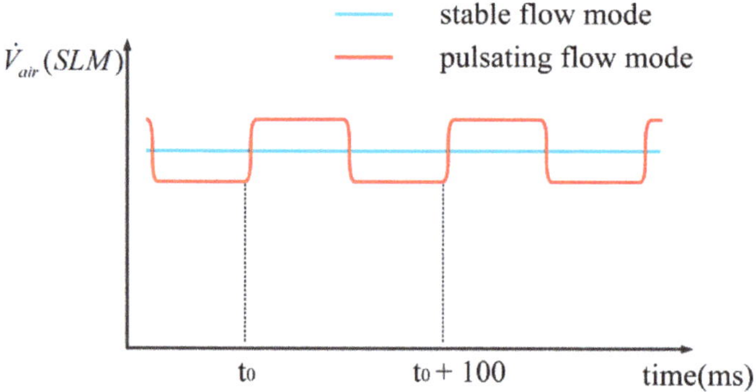

Figure 3. Air flow rates in different modes.

2.2. The Multi-Burner and Arrangement of Electrodes

The multi-burner consisted of an upper cover plate, three stainless pipe bodies with inner diameters of 20 mm, three swirlers, and ceramic tubes implanted with tungsten electrodes, as shown in Figure 4.

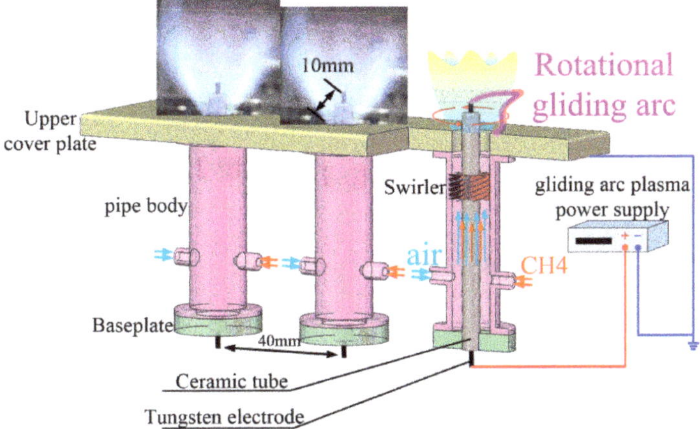

Figure 4. The multi-burner and electrodes.

The tungsten electrode was connected to the high voltage output of gliding arc power supply, while the upper cover plate of the burner was grounded as the cathode. The discharge gap between the top of tungsten electrode and the edge of the burner outlet was 10 mm. When the voltage between the electrode and the cover plate was higher than the breakdown voltage, the discharge occurred and the arc plasma filament was generated. Then the plasma filament was blown out by swirl flow to form the rotational gliding arc.

The strength of the gas swirl was reflected by the swirl number (Sw), which was calculated as Formula (4) and was defined in [24]:

$$Sw = \frac{2}{3}\tan\alpha\frac{1-(R_{in}/R_{ex})^3}{1-(R_{in}/R_{ex})^2} \tag{4}$$

where α is the angle between swirler vane and axis of pipe body, R_{ex} is the external diameter of swirler, and R_{in} is the inner diameter of the swirler. Usually a stable recirculation zone can be generated when Sw is higher than 0.6 (known as strong swirl). The recirculation zone disappeared when Sw approached 0.4 (known as weak swirl) and the flame remained stable because of the shear of the vortex [25]. Three different angles of swirler vanes (30°, 45°, and 60°) were used in the experiment, and the corresponding swirl numbers were: $Sw(30°) = 0.449$, $Sw(45°) = 0.778$, $Sw(60°) = 1.347$, respectively. The extinction performances of both strong and weak swirl were tested in the experiment.

2.3. The Gliding Arc Plasma Power Supply and Measurements

The gliding arc plasma was generated by a low-temperature plasma power supply (CTP2000K). The waveform was a modulated sine wave, and the output voltage varied from 0 to 30 kV. The air and methane came from the compressed tanks, and were controlled by D08-1 mass flow controllers (MFCs). The pressure of the intake flow was no more than 0.3 MPa, and the measurement range of the MFC was 0–200 SLM, with ±2% full scale (F.S.) accuracy and 1–4 s response times.

The current and voltage were measured simultaneously using a Tektronix TCP0030A current probe (120 MHz bandwidth and 0–30 A range) and a P6015A high voltage probe, and were then recorded by the MDO3024 oscilloscope.

The pressure was captured by ICP® PCB pressure sensor with sampling frequency of 1 kHz. The rise time was shorter than 1.0 µsec and non-linearity within 1.0% F.S.

The evolution of CH luminescence in pulsating flame and the stretch of the gliding arc plasma were imaged using a Phantom v251112 high-speed camera equipped with a f/1.8 Nikon zoom lens; a bandpass filter (centered at 430 nm, with a 10-nm full width at half maxima) was also adopted.

3. Results and Discussion

3.1. Characteristics of Gliding arc Plasma Discharge in Stable Flow Mode

Many studies have confirmed that the air flow rates have a great influence on gliding arc plasma [26,27]. In this experiment, the current, voltage, and power waveforms were measured with different flow rates. At the same time, the images of gliding arc plasma topology were recorded simultaneously. During the test process, the supply of CH_4 was cut off, and the pulsating air flow generator was in stable flow mode, which meant the combustor operated in stable cold flow conditions. The characteristics of gliding arc plasma with two specific air flow rates ($(\overline{\dot{V}_{air}})_{stable} = 30SLM$, $(\overline{\dot{V}_{air}})_{stable} = 120SLM$) were analyzed specifically and different discharge types were investigated.

The t_a was defined as the moment that a new arc was generated, while the t_b was defined as the moment that this new arc broke up. When $(\overline{\dot{V}_{air}})_{stable} = 30SLM$ it could be seen that the triangle-shaped voltage waves of gliding arc plasma appeared approximately periodically (Figure 5). The evolution of gliding arc topology captured by high speed camera is presented in Figure 6. At moment t_a (t_a = 18.533 ms) the former arc began to extinguish and a new arc formed. Meanwhile, the voltage amplitude decreased from 6.8 kV to 2 kV and the initial arc discharge power was then only 200 W. As the arc elongated gradually and was blown out of the combustor by the swirl airflow (Figure 6), the amplitudes of voltage and power increased, but the current amplitude basically remained in the order of hundreds of milliamperes (Figure 5b). With the increase in arc length, the power for maintaining the arc channel also increased. Once it exceeded threshold of power supply, the arc broke up and the plasma filament was extinguished, as shown in the image at moment t_b in Figure 6. The peak voltage was about 4.5 kV at moment t_b, and then decreased sharply. Instead of the arc plasma, which disappeared, a new arc was generated in another position randomly because of the swirl flow. As the process was repeated periodically, the gliding arc discharge was maintained.

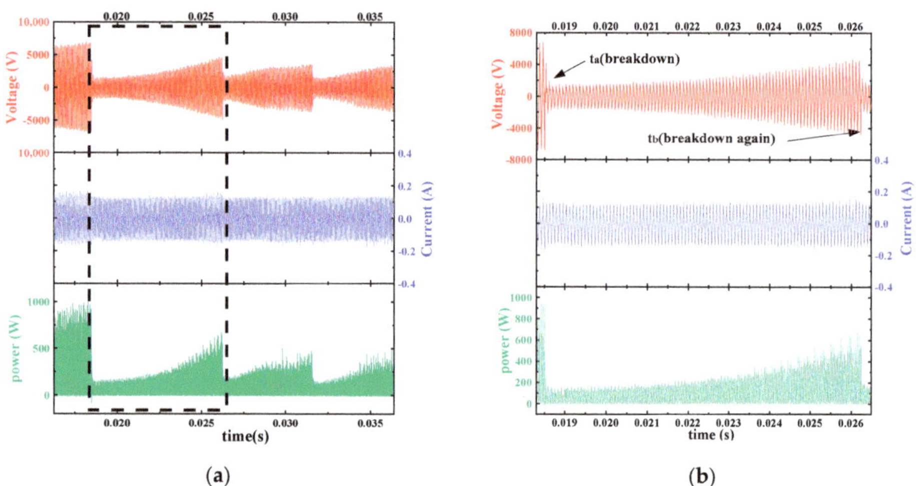

Figure 5. (**a**) Voltage, current, and power waveforms of the gliding arc discharge when average flow rate in stable mode was 30SLM. (**b**) Zoomed-in view voltage, current, and power waveforms (from t = 0.0183 s to t = 0.0265 s).

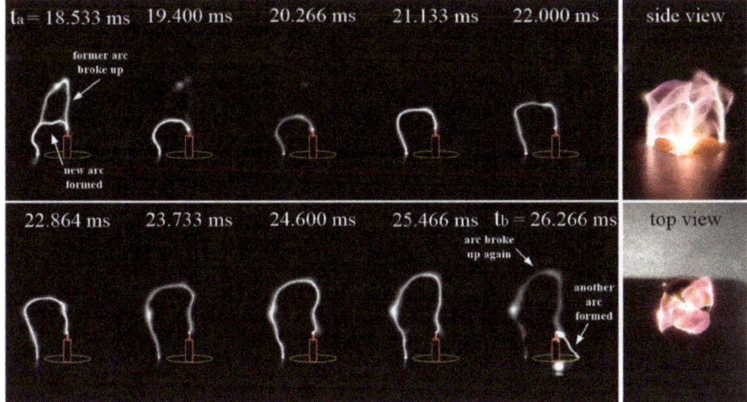

Figure 6. Simultaneous images of gliding arc topologies captured by a high-speed camera (sample rate: 30,000 fps, exposure time: 29.23 μs) and two pictures taken using a regular digital camera when average flow rate in stable mode was 30SLM.

Figures 5 and 6 illustrate that the period time of gliding arc from formation to fracture was 7 ms in approximately $(\overline{\dot{V}}_{air})_{stable} = 30SLM$ air flow, and the voltage waveform in the experiment was similar that in [28].

The t_a was defined as the moment that a new arc was generated, the t_b was defined as the moment that the peak current happened, and the t_c was defined as the moment that this new arc broke up. The characteristics of discharge in higher flow rate are illustrated in Figure 7. When $(\overline{\dot{V}}_{air})_{stable} = 120SLM$, the period of arc plasma discharge was significantly shortened, and the form of the arc also differed greatly from the arc in $(\overline{\dot{V}}_{air})_{stable} = 30SLM$. Similarly, the evolution of gliding arc topology is presented in Figure 8. As shown in Figures 7b and 8, a new arc was generated at moment t_a and the voltage amplitude was close to 2 kV. Then the amplitude of voltage increased from t_a to t_c and the plasma filament elongated simultaneously. This new arc extinguished eventually at moment t_c, at which the voltage reached 8.4 kV and was almost two times higher than that in $(\overline{\dot{V}}_{air})_{stable} = 30SLM$ conditions. The peak of transient discharge power at moment t_c in $(\overline{\dot{V}}_{air})_{stable} = 120SLM$ was larger than that in $(\overline{\dot{V}}_{air})_{stable} = 30SLM$ as well. The discharge process was maintained at 1.2 ms, a shorter period than that in $(\overline{\dot{V}}_{air})_{stable} = 30SLM$, indicating that the discharge period decreased as the air flow rate increased. In particular, at moment t_b the emission of plasma filament suddenly became much brighter, coupled with the obviously transient current peak that was up to 1.1 A in 1 μs. The phenomenon was very different from the ordinary discharge and was denominated as a spark-type discharge. Correspondingly, the ordinary discharge, which was characterized by a long duration and an amplitude of hundreds of milliamperes in the currents (like the currents in Figure 5a), was recognized as a glow-type discharge. The experimental result was in accordance with [27,29].

From the discussion, the characteristics of gliding arc discharges in low flow rate and high flow rate can be summarized with respect to:

(1) The amplitudes of voltages in discharge processes. In high flow rate conditions, the voltage reached over 5 kV when the former arc broke up and the new arc formed, higher than the voltage found in the low flow rate conditions (about 4 kV). The amplitude of transient power with a high flow rate was also larger than that with a low flow rate.

(2) The periods of discharge processes. The period of gliding arc from forming to extinguishing was about less than 1 ms in high flow rate, while the period in low flow rate was approximately 7 ms. The period of discharge processes decreased as the flow rate increased.

(3) The discharge types. With the low flow rate, the discharge always maintained the glow type discharge and no current peak formed. However, with a high flow rate, another discharge type known as the spark-type began to appear, which generated many current peaks. When the flow rate increased, the possibility of conversion from glow type to spark type increased as well.

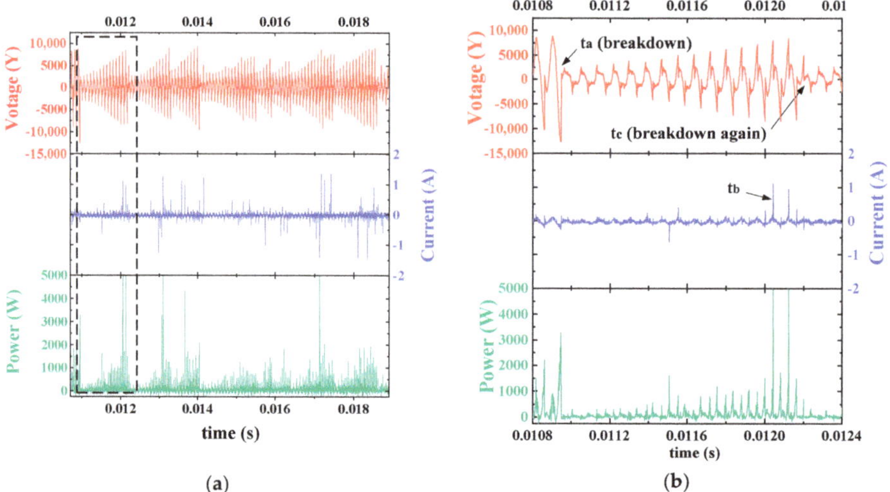

Figure 7. (**a**) Voltage, current, and power waveforms of the gliding arc discharge when average flow rate in stable mode was 120SLM. (**b**) Zoomed-in view voltage, current, and power waveforms (from t = 0.0108 s to t = 0.0124 s).

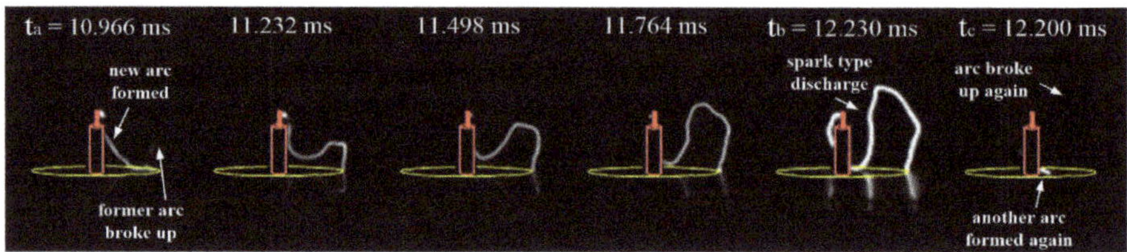

Figure 8. Simultaneous images of gliding arc topologies captured by a high-speed camera when average flow rate in stable mode was 120SLM (sample rate: 30,000 fps, exposure time: 29.23 μs).

Figure 9 analyzes the variation of average power of the gliding arc discharge with different $\overline{\dot{V}}_{air}$ values in stable and pulsating flow mode. It was found that with the increase in air flow rate the average power decreased slightly. The average power in Figure 9b was slightly lower than that in Figure 9a, but the two curves were similar, indicating that compared with the flow rate, the flow mode seemed to have little influence on discharge. Together with Figures 5 and 7, it could be concluded that although the peak of transient power in the $(\overline{\dot{V}}_{air})_{stable} = 120SLM$ condition was larger than that in the $(\overline{\dot{V}}_{air})_{stable} = 30SLM$ condition, the average power was smaller. The reason was probably that the discharge period time decreased as the air flow rate increased, so the high transient power in high flow rate could not be maintained for a long time and quickly decreased, and the average power decreased accordingly.

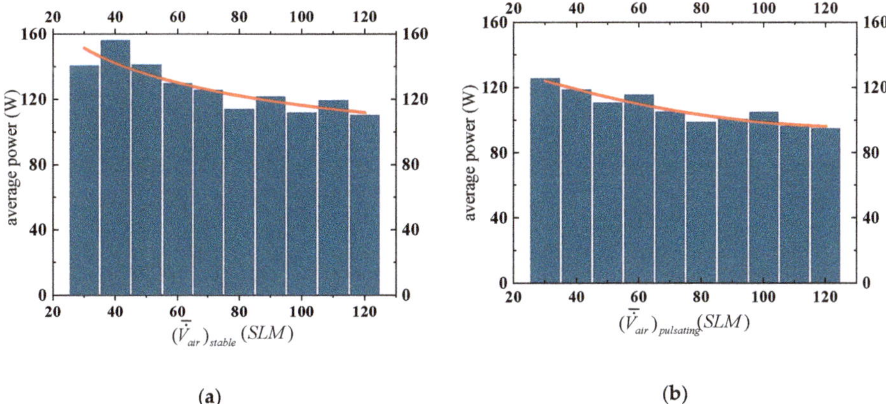

Figure 9. (**a**) The average power in stable flow mode; (**b**) The average power in pulsating flow mode.

3.2. The Influence of Static Instability on the Lean Blowout Limits of Swirl Flame

In the experiment, the lean blowout limits of swirl flame were detected. The average equivalence ratios (represented by $\overline{\varphi}$) of blowout were measured in different flow modes, different $\overline{\dot{V}}_{air}$, and different swirl numbers (represented by Sw). Then these measured average equivalence ratios were connected to form the lean blowout limits.

Figure 10 shows the lean blowout limits of swirl flames in different flow mode conditions. When the Sw and flow modes were fixed and the $\overline{\dot{V}}_{air}$ increased, the $\overline{\varphi}$ of blowout increased as well, which indicated that the extinction performance of swirl flame deteriorated as the air flow rate increased.

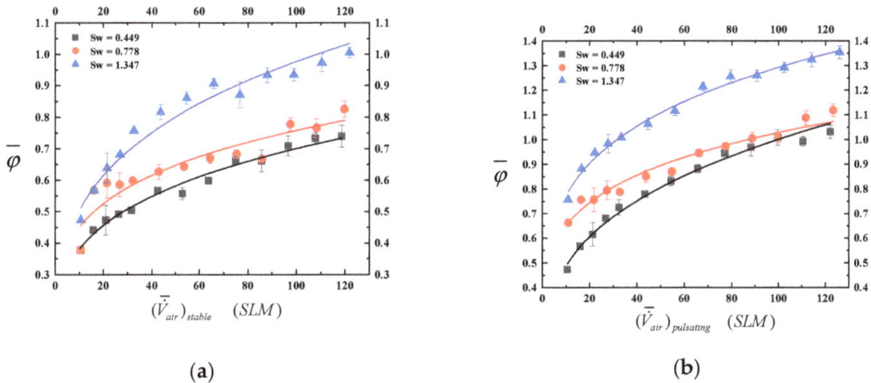

Figure 10. (**a**) The lean blowout limits in stable flow mode; (**b**) The lean blowout limits in pulsating flow mode.

With the flow mode unchanged, when the Sw increased from 0.449 to 1.347, the lean blowout limit gradually increased correspondingly. In other words, the lean blowout limit worsened as the Sw increased, similar to [25].

When the Sw was fixed and the flow mode changed from stable flow to pulsating flow, the static flame instability was generated, and the lean blowout limits became higher than in stable flow mode.

With three different Sw from 0.449 to 1.347, the relative increase in equivalence ratio (represented by $(\Delta\overline{\varphi})_{relative}$) in pulsating flow mode was calculated by Formula (5), and is depicted in Figure 11.

$$(\Delta\overline{\varphi})_{relative} = \frac{\overline{\varphi}_{pulsating} - \overline{\varphi}_{stable}}{\overline{\varphi}_{stable}} \tag{5}$$

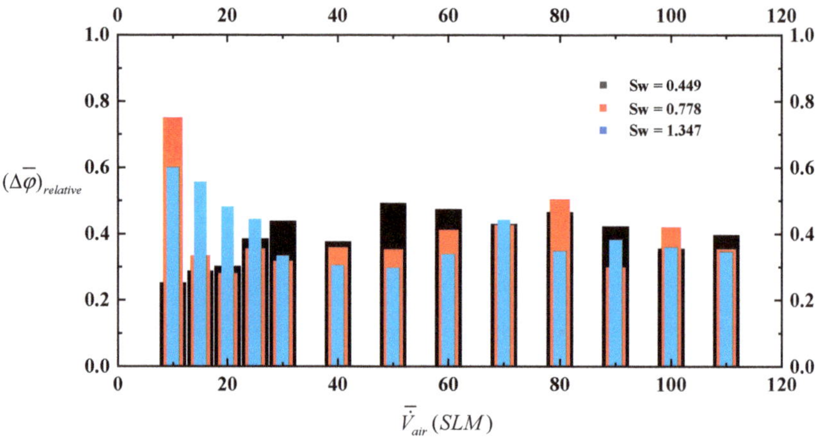

Figure 11. The relative increases in equivalence ratio when the flow mode changed from stable to pulsating.

In Figure 11, compared with stable mode, the blowout equivalence ratios in pulsating mode increased relatively by about 20–50%, indicating that the inlet flow oscillation had a negative influence on extinction performance of swirl flame. In addition, no rules could be determined from the variation of $(\Delta\overline{\varphi})_{relative}$ with the increases in Sw or $\dot{\overline{V}}_{air}$. It seemed that the relative increase in equivalence ratio was mainly related to the change of flow modes.

3.3. The Improvement of Lean Blow-Out Limits with Gliding Arc Plasma

Previous experiments [30] found that the flame was extinguished temporarily and then relighted by the burned gas when the equivalence ratio approached lean blowout limits. If the equivalence ratio was to decrease further, unburned gas would not be reignited and the flame would be completely extinguished.

Fortunately, arc plasma can act as an additional heat source, which is believed to be helpful for the re-ignition process. The lean blowout limits of swirl flame in pulsating flow with gilding arc plasma (defined as the plasma-pulsating flow mode) were measured and are shown in Figure 12.

Compared with Figure 10b, the lean blowout limit in Figure 12 reduced significantly with the use of gliding arc plasma, and the relative decreases in $\overline{\varphi}$ (represented by $(\Delta\overline{\varphi})_{relative}$) were calculated using Formula (6) in order to better reflect the advantage of the gliding arc plasma. The $(\Delta\overline{\varphi})_{relative}$ is depicted in Figure 13.

$$(\Delta\overline{\varphi})_{relative} = \frac{\overline{\varphi}_{plasma-pulsating} - \overline{\varphi}_{pulsating}}{\overline{\varphi}_{pulsating}} \tag{6}$$

In Figure 13 the negative value shows that the blowout equivalence ratio in plasma-pulsating flow mode was lower than that in pulsating flow mode. The lean blowout limit generated by the inflow oscillation was expanded by more than 40% with gliding arc plasma. Interestingly, unlike in Figure 11, the trend of the $(\Delta\overline{\varphi})_{relative}$ is obvious with the increase in $\dot{\overline{V}}_{air}$ in Figure 13. When the $\dot{\overline{V}}_{air}$

rose, the $(\Delta\overline{\varphi})_{relative}$ decreased, indicating that the effect of the plasma with the high flow rate was weaker than it was with the low flow rate. Figure 9 shows that the average power of the gliding arc in high flow rate was smaller than that in low flow rate, so it is reasonable that the effect of the plasma was weakened in high flow rate.

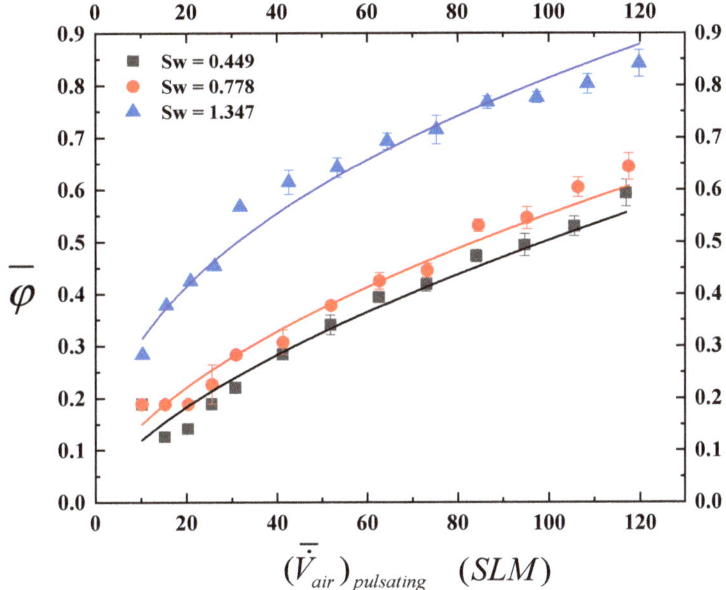

Figure 12. The lean blowout limits in plasma-pulsating flow mode.

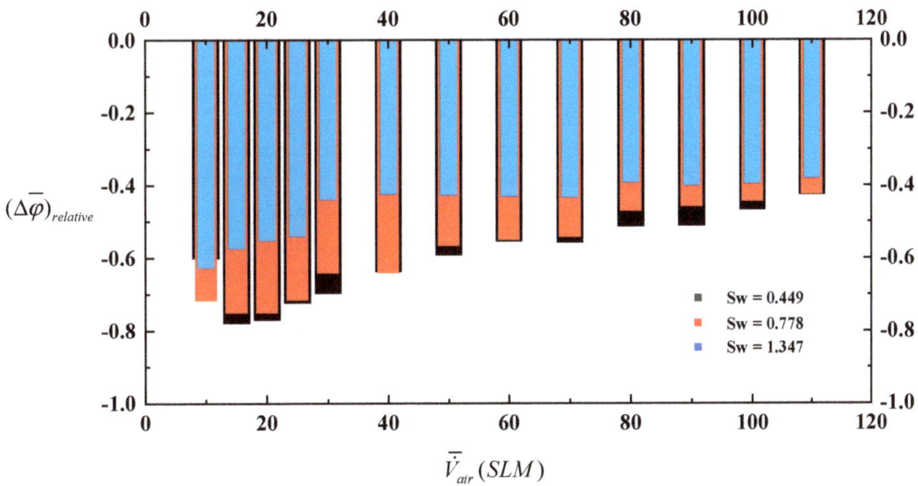

Figure 13. The relative decreases in equivalence ratio when gliding arc plasma was applied to the pulsating flame.

As to why the lean blowout limit could be extended by plasma, one of the reasons could be that the plasma continuously transformed the electric energy into the heat, injected heat into the flame region, and impaired the blowing effect of pulsating airflow on flame heat so the lean blowout limit of pulsating swirl flame was greatly extended. The explanation needs to be confirmed through the dynamic analysis of swirl flame in pulsating flow with plasma.

3.4. Dynamic Analysis of Swirl Flame in Pulsating Flow with Plasma

The pressure in burner outlet was measured and the frequency was calculated in pulsating flow mode, as shown in Figure 14. It can be seen from the pressure curve that the combustion oscillation of swirl flame was excited by the inlet air pulsation, and the main frequency was about 10 Hz. As reported in [4], the frequency of static instability is usually below 30 Hz. It is thus believed that the pulsating flow mode in this experiment showed typical static instability.

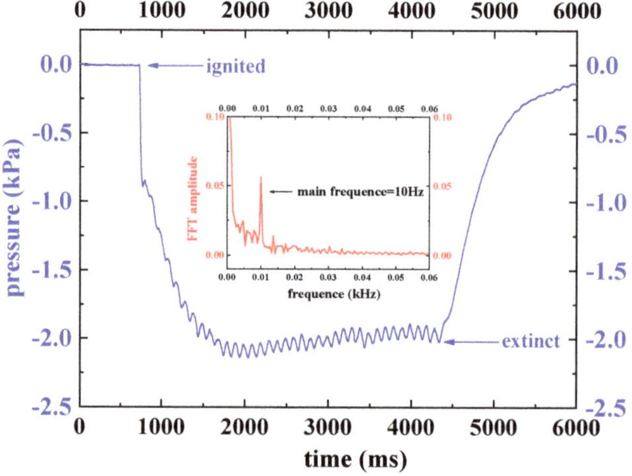

Figure 14. The time signal and frequency signal of flame pressure when $(\overline{\dot{V}}_{air})_{pulsating} = 120SLM$.

The extinction of the swirling flame was a dynamic process consisting of flame front fracture, partial extinction, re-ignition caused by high temperature residual gas, and global extinction [31]. The change of the flow velocity generated by the pulsating air damaged the recirculation zone of swirl flame, causing flame front breaking, combustion instability, and extinction [32].

The dynamic processes of re-ignition and extinction of swirl flames in pulsating air flow mode and plasma-pulsating flow mode are were illustrated in Figure 15. The working conditions include $(\overline{\dot{V}}_{air})_{pulsating} = 120SLM, Sw = 1.347, \overline{\varphi} = 1.34$. In Figure 15a, when the transient air flow rate decreased (from t = 10 ms to t = 60 ms), the flame volume and brightness increased continuously, with both reaching a maximum at t = 60 ms. As the transient flow rate increased gradually from t = 60 ms to t = 100 ms, the flame firstly broke, and then the burned gas left the burner outlet. At t = 100 ms no more swirl flame at the burner outlet could be observed. Compared with Figure 15a, the main difference in Figure 15b was that the swirl flame existed around the gliding arc plasma throughout the whole pulsating period. It can be concluded that the gliding arc plasma acted as an ignition source which made the swirl flame stabilize at the burner outlet.

Generally, the chemiluminescence of radicals like CH, OH, and CH2O is related to the reaction intensity and heat release rate of the flame [33–35]. For example, CH2O is the symbol of the preheat region, OH represents the high-temperature region [34], and CH is regarded as the indicator of the flame front [35]. According to previous research [36–38], CH could better reflect the local or global heat release rate. In the experiment, the high-speed camera with bandpass filter was used to capture the instantaneous structure of CH radical, which could reveal the distribution of heat release rate in both pulsating flow mode and plasma-pulsating flow mode, as shown in Figure 16.

The dynamic processes of CH distribution are shown in Figure 16a,b. The transient flow rate gradually increased during the first 50 ms. As the valve was closed, it decreased in the following 50 ms. It can be seen that the CH mainly concentrated on the top of swirl flame, in accordance with [39]. Within the first 60 ms, there was little difference between Figure 16a,b. However, the CH in Figure 16b was significantly stronger than that in Figure 16a after t = 50 ms. The comparison showed that plasma has a positive effect on the heat release rate, which may be one of the reasons why the lean blowout limit was extended with gliding arc plasma.

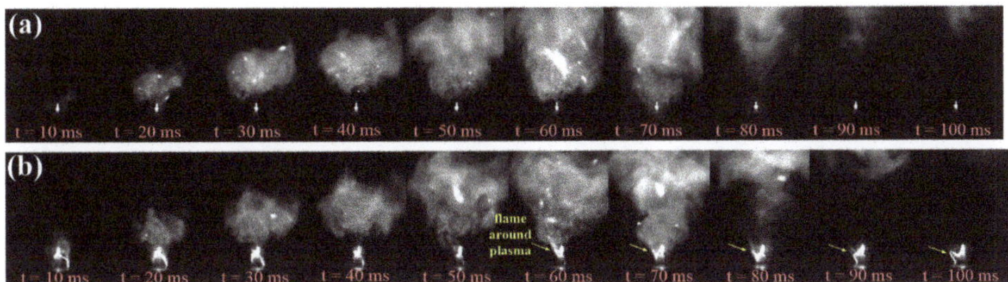

Figure 15. (**a**) The dynamic process of swirl flame in pulsating air flow mode; (**b**) the dynamic process of swirl flame in plasma-pulsating flow mode (sample rate: 30,000 fps, exposure time: 29.23 μs).

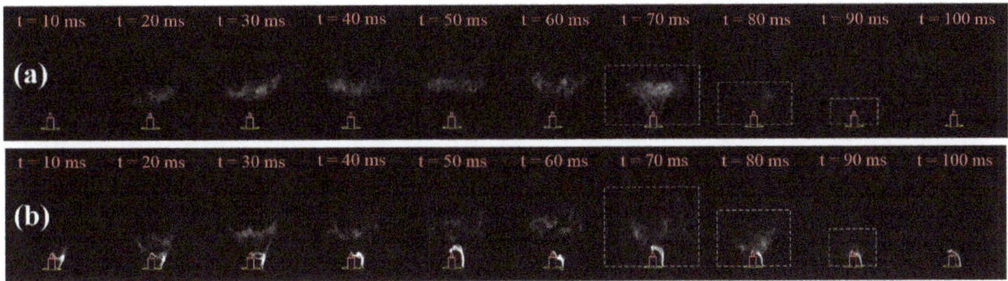

Figure 16. (**a**) The dynamic process of CH distribution in pulsating air flow mode; (**b**) The dynamic process of CH distribution in plasma-pulsating air flow mode (sample rate: 1000 fps, exposure time: 1 ms).

4. Conclusions

The characteristics of gliding arc, as well as its control effect on static instability of premixed swirl flame, were investigated experimentally, and the main conclusions can be summarized as follows:

(1) The gliding arc plasma was characterized by periodic discharge, and its current, voltage, and power waveforms as well as plasma topology were related to the air flow rate. In the low flow rate, the discharge kept the glow type discharge and no current peaks formed. As the flow rate increased, another discharge type named the spark type began to appear, with many current peaks. When the flow rate increased, the period of gliding arc discharge and the average power decreased, while the amplitude of the voltage wave and the peak of instantaneous discharge power increased.

(2) The extinction performance of the flame was influenced deeply by the static flame instability. When the flow mode changed from stable flow to pulsating flow, static flame instability was generated, and the lean blowout limit became higher than in stable flow mode. Apart from the static instability, swirl number also influenced the extinction performance, as the lean blowout limit increased as the swirl number increased.

(3) As the gliding arc plasma was adopted, the lean blowout limit of swirl flame in pulsating flow mode was significantly reduced and was found to be better than the limit of the stable flame. It seemed that when the air flow rate increased, the beneficial effect of the plasma was weakened.

(4) In contrast to the flame dynamic process in pulsating flow mode, the flame existed throughout the whole pulsating period with gliding arc plasma. Thus, it seemed that the gliding arc plasma acted as an ignition source in pulsating mode, which made the swirl flame stabilize at the burner outlet. The CH distribution showed that plasma had a positive effect on the heat release rate, which may be one of the reasons why the lean blowout limit was extended with gliding arc plasma.

In the experiment the discharge characteristics of gliding arc in different swirling flow rates were studied mainly based on the current, voltage, power waveforms, and plasma topology, but other parameters of the gliding arc, such as translational, rotational, and vibrational parameters as well as electron temperatures and the reduced electric field were not researched. Future works should be focused on the measurements of plasma parameters in order to provide a better understanding of the mechanism of plasma assisted-combustion. In addition, the species of the gliding arc plasma in swirl flame also need to be investigated.

Author Contributions: Conceptualization, W.C. (Weiqi Chen) and W.C. (Wei Cui); methodology, D.J.; validation, W.C. (Weiqi Chen); writing—original draft preparation, W.C. (Weiqi Chen); writing—review and editing, D.J. and S.H. All authors have read and agreed to the published version of the manuscript.

Funding: This research was funded by two programs of National Natural Science Foundation of China with grant numbers No. 91641204 and No. 51807204.

Acknowledgments: Thanks to the great support of the National Natural Science Foundation of China.

Conflicts of Interest: The authors declare no conflict of interest.

References

1. Oefelein, J.C.; Yang, V. Comprehensive Review of Liquid-Propellant Combustion Instabilities in F-1 Engines. *J. Propuls. Power* **1993**, *9*, 657–677. [CrossRef]
2. Malbois, P.; Salaun, E.; Vandel, A.; Godard, G.; Cabot, G.; Renou, B.; Boukhalfa, A.M.; Grisch, F. Experimental investigation of aerodynamics and structure of a swirl-stabilized kerosene spray flame with laser diagnostics. *Combust. Flame* **2019**, *205*, 109–122. [CrossRef]
3. Nair, S.; Lieuwen, T. Acoustic detection of blowout in premixed flames. *J. Propuls. Power* **2005**, *21*, 32–39. [CrossRef]
4. Mongia, H.C.; Held, T.; Hsiao, G.; Pandalai, R. Challenges and progress in controlling dynamics in gas turbine combustors. *J. Propuls. Power* **2003**, *19*, 822–829. [CrossRef]
5. Strutt, J.W.; Rayleigh, B. *The Theory of Sound*; Macmillan: London, UK, 1945.
6. Candel, S.; Durox, D.; Schuller, T.; Bourgouin, J.-F.; Moeck, J.P. Dynamics of swirling flames. *Annu. Rev. Fluid Mech.* **2014**, *46*, 147–173. [CrossRef]
7. Ateshkadi, A.; McDonell, V.G.; Samuelsen, G.S. Lean blowout model for a spray-fired swirl-stabilized combustor. *Proc. Combust. Inst.* **2000**, *28*, 1281–1288. [CrossRef]
8. Richards, G.A.; Straub, D.L.; Robey, E.H. Passive control of combustion dynamics in stationary gas turbines. *J. Propuls. Power* **2003**, *19*, 795–810. [CrossRef]
9. Ju, Y.G.; Sun, W.T. Plasma assisted combustion: Dynamics and chemistry. *Prog. Energy Combust. Sci.* **2015**, *48*, 21–83. [CrossRef]
10. Leonov, S.B.; Yarantsev, D.A.; Napartovich, A.P.; Kochetov, I.V. Plasma-assisted chemistry in high-speed flow. *Plasma Sci. Technol.* **2007**, *9*, 760–765. [CrossRef]
11. Leonov, S.B.; Yarantsev, D.A. Plasma-induced ignition and plasma-assisted combustion in high-speed flow. *Plasma Sources Sci. Technol.* **2007**, *16*, 132–138. [CrossRef]
12. Pilla, G.; Galley, D.; Lacoste, D.A.; Lacas, F.; Veynante, D.; Laux, C.O. Stabilization of a turbulent premixed flame using a nanosecond repetitively pulsed plasma. *IEEE Trans. Plasma Sci.* **2006**, *34*, 2471–2477. [CrossRef]
13. Ombrello, T.; Qin, X.; Ju, Y.; Gutsol, A.; Fridman, A. Enhancement of combustion and flame stabilization using stabilized non-equilibrium plasma. In Proceedings of the 43rd AIAA Aerospace Sciences Meeting and Exhibit, Reno, NV, USA, 10–13 January 2005; p. 1194.
14. Ombrello, T.; Qin, X.; Ju, Y.; Gangoli, S.; Gutsol, A.; Fridman, A. Non-equilibrium plasma discharge: Characterization and effect on ignition. In Proceedings of the 44th AIAA Aerospace Sciences Meeting and Exhibit, Reno, NV, USA, 9–12 January 2006; p. 1214.

15. Ombrello, T.; Qin, X.; Ju, Y.; Gutsol, A.; Fridman, A.; Carter, C. Combustion Enhancement via Stabilized Piecewise Nonequilibrium Gliding Arc Plasma Discharge. *AIAA J.* **2006**, *44*, 142–150. [CrossRef]
16. Fridman, A.; Gutsol, A.; Gangoli, S.; Ju, Y.G.; Ombrellol, T. Characteristics of Gliding Arc and Its Application in Combustion Enhancement. *J. Propuls. Power* **2008**, *24*, 1216–1228. [CrossRef]
17. Ombrello, T.; Ju, Y.; Fridman, A. Kinetic Ignition Enhancement of Diffusion Flames by Nonequilibrium Magnetic Gliding Arc Plasma. *AIAA J.* **2008**, *46*, 2424–2433. [CrossRef]
18. Moeck, J.P.; Lacoste, D.A.; Durox, D.; Guiberti, T.F.; Schuller, T.; Laux, C.O. Stabilization of a Methane-Air Swirl Flame by Rotating Nanosecond Spark Discharges. *IEEE Trans. Plasma Sci.* **2014**, *42*, 2412–2413. [CrossRef]
19. Larsson, A.; Zettervall, N.; Hurtig, T.; Nilsson, E.J.K.; Ehn, A.; Petersson, P.; Alden, M.; Larfeldt, J.; Fureby, C. Skeletal Methane-Air Reaction Mechanism for Large Eddy Simulation of Turbulent Microwave-Assisted Combustion. *Energy Fuels* **2017**, *31*, 1904–1926. [CrossRef]
20. Discepoli, G.; Cruccolini, V.; Ricci, F.; Di Giuseppe, A.; Papi, S.; Grimaldi, C.N. Experimental characterisation of the thermal energy released by a Radio-Frequency Corona Igniter in nitrogen and air. *Appl. Energy* **2020**, *263*, 13. [CrossRef]
21. Zhu, J.J.; Ehn, A.; Gao, J.L.; Kong, C.D.; Alden, M.; Salewski, M.; Leipold, F.; Kusano, Y.; Li, Z.S. Translational, rotational, vibrational and electron temperatures of a gliding arc discharge. *Opt. Express* **2017**, *25*, 20243–20257. [CrossRef]
22. Zhu, J.J. Optical Diagnostics of Non-Thermal Plasmas and Plasma-Assisted Combustion. Ph.D. Thesis, Lund University, Lund, Sweden, 2015.
23. Sun, Z.W.; Zhu, J.J.; Li, Z.S.; Aldén, M.; Leipold, F.; Salewski, M.; Kusano, Y. Optical diagnostics of a gliding arc. *Opt. Express* **2013**, *21*, 6028–6044. [CrossRef]
24. Beér, J.M.; Chigier, N.A. *Combustion Aerodynamics*; Applied Science Publication: London, UK, 1972.
25. Cheng, R.K.; Littlejohn, D.; Strakey, P.A.; Sidwell, T. Laboratory investigations of a low-swirl injector with H-2 and CH4 at gas turbine conditions. *Proc. Combust. Inst.* **2009**, *32*, 3001–3009. [CrossRef]
26. Niu, Z.T.; Zhang, C.; Ma, Y.F.; Wang, R.X.; Chen, G.Y.; Yan, P.; Shao, T. Effect of flow rate on the characteristics of repetitive microsecond-pulse gliding discharges. *Acta Phys. Sin.* **2015**, *64*, 8. [CrossRef]
27. Feng, R.; Li, J.; Wu, Y.; Jia, M.; Jin, D.; Lin, B.X. Experimental Research on Discharge Characteristics of Multiple-channel Gliding Arc Discharge. *High Volt. Eng.* **2018**, *44*, 4052–4060.
28. Kong, C.D.; Gao, J.L.; Zhu, J.J.; Ehn, A.; Alden, M.; Li, Z.S. Characteristics of a Gliding Arc Discharge Under the Influence of a Laminar Premixed Flame. *IEEE Trans. Plasma Sci.* **2019**, *47*, 403–409. [CrossRef]
29. Kong, C.D.; Gao, J.L.; Zhu, J.J.; Ehn, A.; Alden, M.; Li, Z.S. Effect of turbulent flow on an atmospheric-pressure AC powered gliding arc discharge. *J. Appl. Phys.* **2018**, *123*, 9. [CrossRef]
30. Nair, S.; Lieuwen, T. Near-blowoff dynamics of a bluff-body stabilized flame. *J. Propuls. Power* **2007**, *23*, 421–427. [CrossRef]
31. Cui, W.; Ren, Y.H.; Li, S.Q. Stabilization of Premixed Swirl Flames Under Flow Pulsations Using Microsecond Pulsed Plasmas. *J. Propuls. Power* **2018**, *35*, 190–200. [CrossRef]
32. Cohen, J.; Bennett, J. An experimental study of the transient flow over a backward-facing step. In Proceedings of the 34th Aerospace Sciences Meeting and Exhibit, Reno, NV, USA, 15–18 January 1996; p. 322.
33. Zhou, B.; Brackmann, C.; Li, Q.; Wang, Z.K.; Petersson, P.; Li, Z.S.; Alden, M.; Bai, X.S. Distributed reactions in highly turbulent premixed methane/air flames Part I. Flame structure characterization. *Combust. Flame* **2015**, *162*, 2937–2953. [CrossRef]
34. Gao, J.; Kong, C.; Zhu, J.; Ehn, A.; Hurtig, T.; Tang, Y.; Chen, S.; Aldén, M.; Li, Z. Visualization of instantaneous structure and dynamics of large-scale turbulent flames stabilized by a gliding arc discharge. *Proc. Combust. Inst.* **2019**, *37*, 5629–5636. [CrossRef]
35. Poinsot, T.J.; Trouve, A.C.; Veynante, D.P.; Candel, S.M.; Esposito, E.J. Vortex-driven acoustically coupled combustion instabilities. *J. Fluid Mech.* **1987**, *177*, 265–292. [CrossRef]
36. Bandaru, R.; Miller, S.; Lee, J.G.; Santavicca, D. Sensors for measuring primary zone equivalence ratio in gas turbine combustors. In Proceedings of the Advanced Sensors and Monitors for Process Industries and the Environment, Boston, MA, USA, 22 January 1999; p. 11.
37. Venkataraman, K.; Preston, L.; Simons, D.; Lee, B.; Lee, J.; Santavicca, D. Mechanism of combustion instability in a lean premixed dump combustor. *J. Propuls. Power* **1999**, *15*, 909–918. [CrossRef]

38. Huang, S.F.; Wu, Y.; Song, H.M.; Zhu, J.J.; Zhang, Z.B.; Song, X.L.; Li, Y.H. Experimental investigation of multichannel plasma igniter in a supersonic model combustor. *Exp. Therm. Fluid Sci.* **2018**, *99*, 315–323. [CrossRef]
39. Liu, Y.; Tan, J.G.; Wang, H.; Gao, Z.W. Radiation Characteristics of Radicals in Low Swirl Methane-Air Flames. *J. Propuls. Technol.* **2019**, *40*, 2022–2029.

Article

MicroscopicCharacteristics and Properties of Fe-Based Amorphous Alloy Compound Reinforced WC-Co-Based Coating via Plasma Spray Welding

Yan Xu [1,2], Yinfeng Wang [3], Yi Xu [3], Mingyong Li [3] and Zheng Hu [3,*]

[1] China Research and Development Academy of Machinery Equipment, Beijing 100089, China; brenxuyan@gmail.com
[2] School of Mechanical and Vehicle Engineering, Beijing Institute of Technology, Beijing 100081, China
[3] Science and Technology on Vehicle Transmission Laboratory, China North Vehicle Research Institute, Beijing 100072, China; wangyf201@gmail.com (Y.W.); xuyitest201@gmail.com (Y.X.); mingyongli80@gmail.com (M.L.)
* Correspondence: huzhengnoveri@gmail.com

Abstract: Plasma spray welding, as one of the material surface strengthening techniques, has the advantages of low alloy material consumption, high mechanical properties and good coating compactness. Here, the Co alloy, WC and Fe-based amorphous alloy composite coating is prepared by the plasma spray welding method, and the coating characteristics are investigated. The results indicate that the coatings have a full metallurgical bond in the coating/substrate interface, and the powder composition influences the microstructures and properties of the coating. The hardness of coatings increases with the mass fraction of the Fe-based amorphous alloy. The spray welding layer has a much higher wear resistance compared with the carbon steel, and the Fe-20 exhibits a superior wear resistance when compared to others. The results indicate that the amorphous alloy powders are an effective additive to prepare the coating by plasma spray welding for improving the wear resistance of the coating.

Keywords: plasma spray welding; WC-Co coating; Fe-based amorphous alloy; microstructure and properties

Citation: Xu, Y.; Wang, Y.; Xu, Y.; Li, M.; Hu, Z. Microscopic Characteristics and Properties of Fe-Based Amorphous Alloy Compound Reinforced WC-Co-Based Coating via Plasma Spray Welding. *Processes* **2021**, *9*, 6. https://dx.doi.org/10.3390/pr9010006

Received: 16 November 2020
Accepted: 21 December 2020
Published: 22 December 2020

1. Introduction

The rapid development of the industry has led to increasingly higher requirements for the performance and service life of various mechanical parts [1–3]. Wear and corrosion are the prime reasons for material failure during use, which seriously affect the performance quality and service life of mechanical products [4–7]. Therefore, improving the wear resistance of the material surface is of great significance to reduce costs and save resources [8,9]. In order to improve the service life of structural steel in harsh environments, the plasma spray welding is an effective method to modify and strengthen the surface of the steel to form a new alloy coating to improve the wear and corrosion resistance of the material and reduce the production cost [10–13].

In the plasma spray welding technique, suitable alloy powders are selected according to the actual conditions and requirements of surface properties [14–16]. A spray welding layer of cobalt-based alloy has an excellent high temperature impact wear resistance and outstanding high temperature oxidative stability [17,18]. By optimizing the plasma spray welding process parameters, the quality of the spray welding layer has been improved, and different methods were proposed to reduce the spray welding layer cracks [19–22]. By adding different materials to the alloy powder to make composite materials, the structure and performance of the plasma spray welding layer are improved [23,24]. Especially in the aspects of hardness, wear resistance, corrosion resistance and oxidation resistance, the spray welding layer has achieved excellent results [25,26].

Numerous studies have shown that a transition metal-based amorphous alloy can be used directly as alloy powder or added to a nickel base, cobalt base, iron base and other self-fluxing alloyed powders owing to the high melting point, high hardness and good abrasive resistance. For example, $Fe_{45}Cr_{16}Mo_{18}C_{18}B_5$ amorphous coatings were deposited on the mild substrate by plasma spraying, and their corrosion in 3.5 wt.% NaCl solution was compared with those of the substrate and 316 stainless steel. The obtained results imply that the coating has a promising prospect for industrial applications [27]. Moreover, it has been reported that a WC-CoCr/35 wt.% amorphous Fe-based alloy composite coating was prepared by high velocity oxygen fuel spray, and, in comparison with the WC-CoCr coating, the WC-CoCr/35 wt.% amorphous Fe-based alloy composite coating exhibits excellent thermal stability and better corrosion resistance [28].

In this work, WC, Co and Fe-based alloy composite powders are selected to prepare a reinforced coating on the surface of 42CrMo steel using the plasma spray welding method. The Co-based amorphous alloy had excellent physicochemical properties, and the mechanical performance under high-temperature conditions was excellent [22,29]. The WC coating compounded with Co element has a high hardness value in a certain temperature range. At the same time, its impact resistance and toughness are very good, and the porosity of the coating is low, and it is combined with the substrate's high strength [3,30,31]. The influence and mechanism of the added Fe-based amorphous powder amount on the microstructure and properties of the spray welding layer are discussed. The results clearly show that the hardness of coatings increases with the mass fraction of the Fe-based amorphous alloy. The spray welding layer has a much higher wear resistance compared with the carbon steel, and the Fe-20 exhibits a superior wear resistance when compared to others.

2. Materials and Methods

2.1. Experimental Materials

42CrMo steel was adopted as the base material of plasma spray welding, and the 42CrMo chemical composition is shown in Table 1. WC (53–90 μm, 99.9 wt.%), Co alloy (53–165 μm) and Fe-based amorphous alloy (16–54 μm) are all products by Aidun Alloy Material Co., Ltd., Suzhou, China, and their chemical compositions are shown in Table 1.

Table 1. Chemical composition (wt.%) of the raw material.

Powder	C	Si	Mn	Cr	Mo	Ni	P	S	W	B	Fe	Co
42CrMo	0.45	0.37	0.8	1.2	0.25	0.03	0.03	0.03	–	–	rest	–
Co-Based Alloy	3.2	1.0	1.0	26.0	–	22.5	–	–	5.0	–	3.0	rest
Fe-Based Amorphous Alloy	1.7	1.8	–	4.0	–	7.5	–	–	–	3.0	rest	–

2.2. Experimental Methods

Firstly, Co alloy, WC and Fe-based amorphous alloy powders are mixed with ball-milling in a certain ratio (20%WC + Co + 5%Fe, 20%WC + Co + 10%Fe, 20%WC + Co + 15%Fe and 20%WC + Co + 20%Fe marked as Fe-5, Fe-10, Fe-15 and Fe-20). Secondly, the 42CrMo steel was cut into a sample of 30 × 50 × 10 mm using a wire cutting machine. Before plasma spray welding, the sample was polished with coarse sandpaper to remove the oxide layer and adjust the roughness, and then cleaned ultrasonically for 30 min in a bath of acetone to remove the rust, organic and fatty contaminants. Finally, a plasma spray welding was used on the 42CrMo steel surface to prepare the plasma welding layer, and the layer was about 23 × 35 mm. The spray welding layer was arc-shaped and the thickness about 2.7 mm. The technological parameters of the plasma spray welding are shown in Table 2.

Table 2. The plasma spray welding parameters.

Delivering Powder Gas Flow Rate (mL/min)	Plasma Gas Flow Rate (mL/min)	Feeding Voltage (V)	Standoff Distance (mm)	Powder Delivery Amount (g/min)	Current (A)	Bead Diameter (mm)
400	300	20	10	20	125	2

2.3. Detection and Analysis

Microstructure observations and a chemical analysis of the plasma spray welding layer were carried out by a field-emission scanning electron microscope (SEM, S-3000N, Hitachi, Tokyo, Japan) equipped with an energy-dispersive X-ray spectrometer (EDS). The phase composition of the sample was analyzed by the X-ray diffraction (XRD, D8 ADVANCE, BRUKER, Karlsruhe, Germany) and the parameters were set as follows: D/Max 2500PC, Cu Kα radiation, λ = 0.15418 nm, scan rate is 4°/min. Furthermore, the micro hardness of the sample was tested by adopting a micro-sclerometer (MH-3, Minks Testing Equipment Co., Ltd., Xi'an, China) for 10 s, with an applied load of 500g, and the test parameters were determined according to the composition of the material, the uniformity of the structure, the size and the surface roughness [32]. The wear resistance of the sample was tested by adopting an abrasive wear testing machine (ML-100, Jingcheng Testing Technology Co., Ltd., Jinan, China), with an applied load of 10 N. The stroke was 121.0 m and wear materials 600#SiC waterproof abrasive paper were used [33].

3. Result and Discussion

The results of the XRD analysis of the 20%WC + Co spray welding layer with different contents of Fe-based amorphous alloy are shown in Figure 1. It was found that the spray welding layer is mainly composed of carbides such as WC, Cr,Fe,Ni,Co(W,C) and $(Cr,Fe,Ni,Co)_3W_3C$, as well as the alloy phase of Cr, Fe, Ni and Co. Cr,Fe,Ni,Co(W,C) is an fcc solid solution of W and C in the metal lattice with a larger lattice parameter, whereas $(Cr,Fe,Ni,Co)_3W_3C$ is a metastable sub-stoichiometric ternary carbide called η-phase (M_6C, where M = metal). Since the Cr, Fe and Co elements have a strong affinity with the C element, they could form carbides easily. The carbides have both a high hardness and excellent wear resistance and can be used as a hard phase supporting matrix, which can greatly improve the hardness and wear resistance of the spray welding layer [34]. Figure 1b shows the XRD analysis results of Fe-20. Compared to Fe-5, a much harder phase was formed on the surface of the substrate. This can be attributed to the fact that an increase of the Fe content would generate a harder phase in the spray welding process, which is conducive to comprehensively improving the hardness, wear resistance and other mechanical properties of the spray-welding layer.

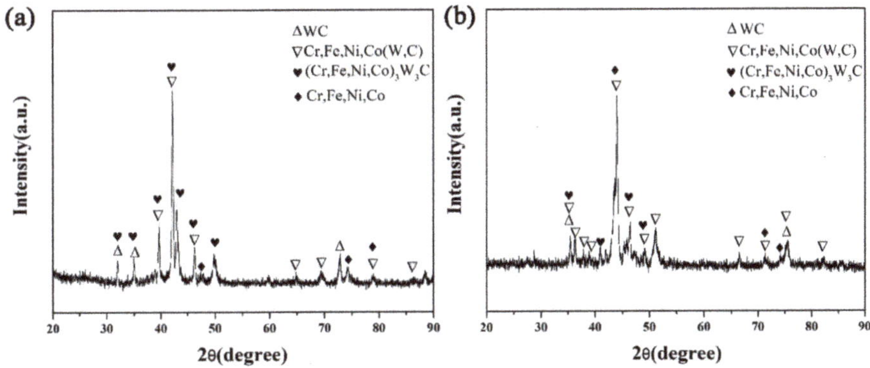

Figure 1. XRD patterns of the coatings with different Fe-based amorphous alloy contents. (a) 5%, (b) 20%.

The typical surface morphology of the coatings with a different Fe-based amorphous alloy content are characterized by SEM, as shown in Figure 2. The Fe-based amorphous alloy content of 5% is shown in Figure 2a, where it can be clearly seen that the white phase is the carbide phase, which is the main existing style of carbide particles in the spray welding layer, and the dark gray phase is mainly the matrix solid solution phase. As shown in Figure 2, the white carbide phase mainly exists in the form of regular dendrites in the spray welding layer. As the Fe-based amorphous alloy content increases, the number and size of dendritic solidification structures also increase, as shown in Figure 2b. It is worth noting that few WC particles are observed in the coating surface, which is due to the different melting points and densities of WC and the metal phase during the rapid cooling rate in the plasma spray welding process. The WC melting point, which is the highest compared to that of the metal phase, is about 2870 °C. During the process of coating formation, the metal phase is melted into liquid owing to the higher hot plasma energy, while the WC powder is hard to melt. Then, the WC powder with a high density would sink down inward because of gravity. In order to confirm the main phase in the coating surface, the elemental composition of the surface of the spray welding layer was analyzed, and the results of points 1–4 are shown in Table 3. It can be clearly seen that points 1 and 3 have the same elements, including C, Si, Cr, Fe, Co, W and Ni. And so do points 2 and 4, compared with points 1 and 3, which lack the Si element. The results indicate that the white area's chemical component in the dendritic crystal is composed of WC and that the dendritic crystal is mainly composed of $(Cr,Fe,Ni,Co)_3W_3C$, while the matrix's solid solution phase is mainly composed of Cr,Fe,Co,Ni(C). Moreover, it was found that the Fe, Cr, Ni and Co elements in the carbide phase are significantly less than that in the matrix, while the W elements in the carbide phase are more numerous than that in the matrix.

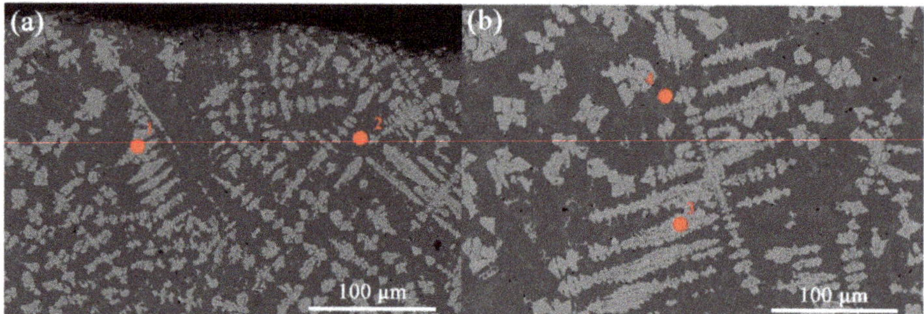

Figure 2. Surface SEM images of different coatings: (**a**) Fe-5 and (**b**) Fe-20.

Table 3. Chemical composition (at %) of the surface of the spray welding layer.

	C	Si	Cr	Fe	Co	W	Ni
1	16.96	1.76	15.69	7.73	21.22	25.56	11.08
2	15.07	–	22.42	8.98	36.09	2.21	15.23
3	19.21	2.93	14.94	9.59	20.79	22.96	9.58
4	17.74	–	19.7	10.61	35.15	2.05	14.75

Figure 3 shows the SEM of the cross-section of the samples' spray welding layer. The result shows that the coating has a dense microstructure and the thermal plasma beam can ensure metallurgical bonding between the coating and the 42CrMo steel. As shown in Figure 3a,e, it was found that the WC particles are not evenly distributed. There are few WC particles in the surface zone of coating, while the WC particles are concentrated in the fusion zone nearing the steel surface. The microstructures of the surface zone are shown in Figure 3b,f. It is clear that most of the carbide phases are dendritic structures, some of which are irregular in shape and distributed among the dendritic structures. Figure 3c,g show the microstructures of the middle zone. Compared with the surface zone, the irregular WC particles in the middle zone increase and the carbide phases with dendritic structure decrease. Figure 3d,h shows the structure of the fusion zone. It can be clearly seen that there are plenty of WC particles with a surrounded dendritic structure. This is due to the plasma spray welding process. The melting point of WC is too high, and the WC exists in a semi-melted form in the liquid metal. The semi-melted WC particles can be used as nucleation particles in the liquid phase solidification process to promote non-uniform nucleation, causing the local segregation of alloy components and promoting the growth of nucleation particles in the form of dendrites during the solidification of the liquid phase metal [35]. In a certain crystal grain range, the hardness decreases with the increase of WC crystal grains. This is due to the small size effect of fine grain WC, large surface energy, effective flow for liquid phase diffusion during the spray welding process, uniform grain growth and high densification, which improves the mechanical properties [36,37]. As shown in Figure 3c,d, it was found that the crystal grain of WC is bigger than tin Figure 3g,h indicating that the properties of Fe-20 are superior to those of Fe-5.When the grains have an uneven distribution and an irregular growth, the mechanical properties of the coating are reduced. It can be seen that the grains' uneven distribution in Fe-5 and Fe-20 affect the performance of the coating. Besides, the shape of the carbide phase is significantly different from the difference in local alloy composition carbides and temperature gradient. Furthermore, the EDS analysis results of the middle zone in Fe-20 are shown in Figure 3i–l. It is obvious that the W elements are mainly distributed in the WC phase and the carbide phase, while the Fe, Cr, Ni and Co elements in the matrix are more numerous than the carbide phase. Several points are selected to further explore the chemical composition in the fusion zone, and the results of the electronic probe are shown in Table 4. It was found that point 1 only contains the C and W elements, indicating that the place of the point is WC. Points 2 and 3 contain the same elements, including C, Si, Cr, Fe, Co, W and Ni. It is confirmed that the WC particles are surrounded by a carbide phase with a dendritic structure. However, the distribution of elements in the carbide phase is quite different at different positions. The carbide phase near the WC phase has more W elements than those away from the WC phase. The reason is that, during the plasma spray welding process, WC gradually melts and releases W elements which would induce a high concentration of W around the WC particles, while these W elements would form a $(Cr,Fe,Ni,Co)_3W_3C$ phase. In addition, the results of EDS component analysis show that the Fe content in point 3 is significantly higher than that in point 2. This is because the melting of the sub-matrix will cause the elements in the matrix to enter the molten pool. As a result, the Fe content is much higher, while the other elements in the fusion zone are diluted.

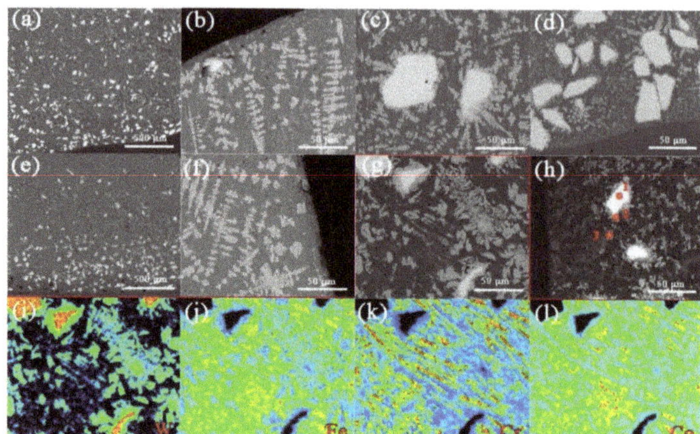

Figure 3. SEM images of (**a–d**) the cross-section of the Fe-5 coating, (**e–h**) SEM images of the cross-section of the Fe-20 coating and (**i–l**) EDS images of the middle zone in the Fe-20 coating.

Table 4. Chemical composition (at %) of the points in the fusion zone.

	C	Si	Cr	Fe	Co	W	Ni
1	45.62	–	–	–	–	54.38	–
2	17.31	2.03	14.24	8.29	19.4	28.52	10.21
3	15.25	1.26	12.02	26.59	18.71	17.69	8.48

The influence of the contents of Fe-based amorphous alloys on the binding quality, microhardness and wear resistance property of the spray welding layer was further studied. Figure 4 shows the binding quality of the coating to the substrate. It can be seen that when the Fe-based amorphous alloy content is 20%, the spray welding layer can be well combined with the substrate, and when the Fe-based amorphous alloy content is relatively lower, the spray welding layer cannot form an excellent cladding with the substrate. When the content of the Fe-based amorphous alloy is 5%, the spray welding current is relatively too large to meet the requirement of the spray welding, so that the spray welding layer cannot be well fused with the substrate, resulting in spray welding defects. The microhardness distribution ranging from substrate to coating is shown in Figure 5a. It can be clearly seen that the hardness of the 42CrMo steel substrate is about 320 HV and the hardness of the spray welding layer is about 650–750 HV. The microhardness of the coating is about two times higher than that of the substrate for each coating. The microhardness of the spray welding layer fluctuates slowly, which may be caused by the difference in the size and quantity distribution of the hard phase. It was found that the hardness gradient in the bonding domain is large, which is a typical characteristic of the hardness distribution of the spray welding layer. This is the result of the sprayed liquid alloy and the matrix, which diffuse and penetrate each other, and the C element in the alloy power plays a solid solution strengthening effect [38,39]. Besides, as the content of Fe-based amorphous alloys increases, the microhardness of the spray welding layer increases, and has a great relationship with the microstructure. The increase of the carbide phase in the spray welding layer and the distribution lead to an increase in the hardness of the spray welding layer [40–42].

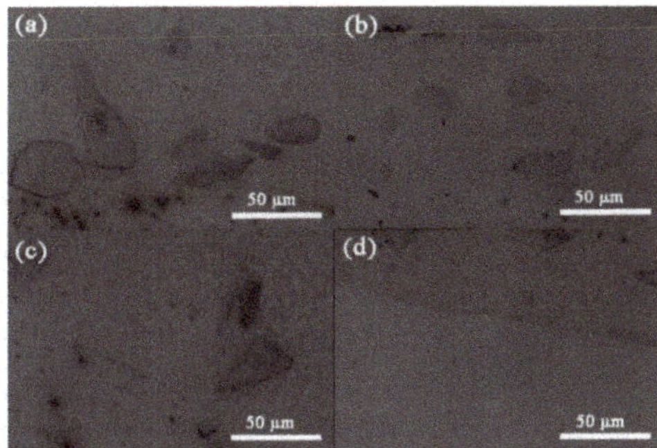

Figure 4. The different contents of the Fe-based amorphous alloy of the binding quality to the substrate (**a**) Fe-5 (**b**) Fe-10 (**c**) Fe-15 (**d**) Fe-20.

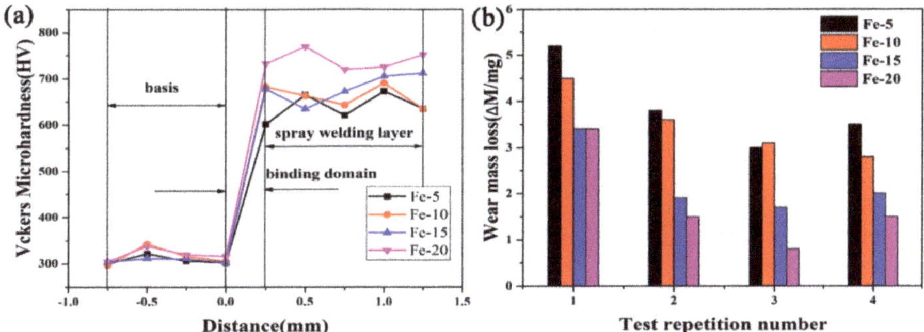

Figure 5. The influence of the Fe-based amorphous alloy content on the mechanical properties of the plasma spray welding layer: (**a**) microhardness and (**b**) wear mass loss.

Wear mass loss is an important factor in characterizing the wear resistance of a material. The results of the wear mass loss are shown in Figure 5b. The wear mass loss of the spray welding layer of Fe-5, Fe-10 and Fe-15 is more than that of Fe-20. The smaller the wear loss rate, the better the wear resistance of the spray welding layer. When the amount of the Fe-based amorphous alloy added is 20%, the spray welding layer has better wear resistance. When the amount of the Fe-based amorphous alloy added is 5–15%, the microhardness and wear resistance are reduced. This is because when a small amount of Fe-based amorphous alloy is added, it can only play a solid solution strengthening effect [33,43]. Additionally, as shown in Figure 4b, the wear mass loss of the following three frictions and wear experiments is relatively balanced, while the wear mass loss of the first time is relatively high. This is because the surface is the outermost in the first wear experiment, which has a lower degree of bonding than the inside.

To further explore the influence of different Fe-based amorphous alloys on the wear resistance of the spray welding layer, frication and wear experiments were carried out on the samples. The morphologies of the worn surfaces of the coatings are shown in Figure 6. Figure 6a–c shows the SEM images of Fe-5, Fe-10 and Fe-15. Deep furrows and scratches are clearly visible on the sample surface, caused by the splashing and agglomeration of Fe-based amorphous alloys. The powder fails to effectively bond with the matrix when it is melted, resulting in a decrease in the Fe-based amorphous alloys content in the spray welding layer and an increase in the austenite phase content. The spray welding layer becomes soft, causing the furrows to deepen. Figure 6d shows the wear morphology of the Fe-20 spray welding layer. Fewer scratches and spalling pits were found, and there are almost no deep furrows in the spray welding layer's wear scars. Meanwhile, more hard phases are scattered in the spray welding layer. The structure is small and uniform, and the hardness and toughness reach a good level. Therefore, the spray welding layer has a better wear resistance.

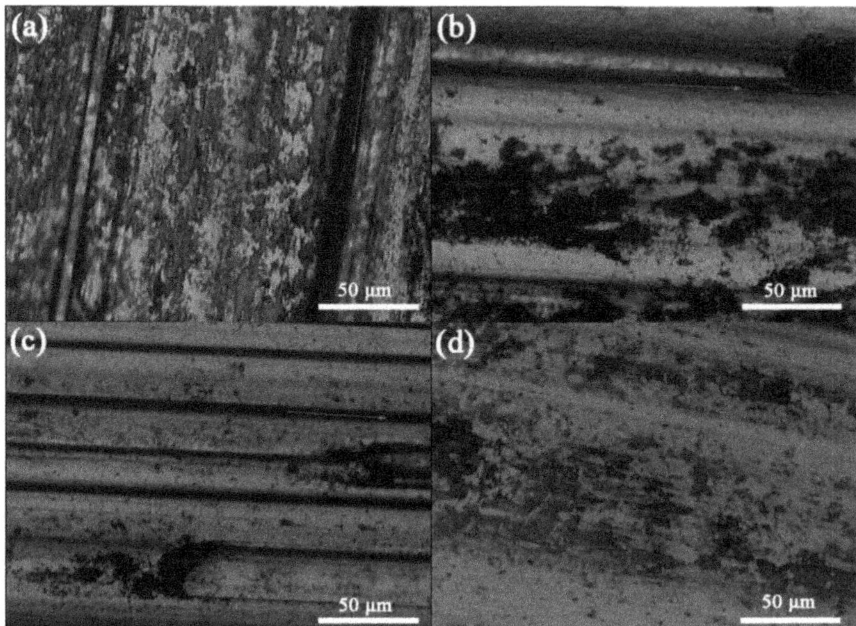

Figure 6. The morphologies of worn coatings: (**a**) Fe-5, (**b**) Fe-10, (**c**) Fe-15 and (**d**) Fe-20.

4. Conclusions

In summary, Fe-based amorphous alloy reinforced WC-Co-based coating was prepared on 42CrMo steel by plasma spray welding. The results indicate that the coatings have a full metallurgical bond in coating/substrate interface and the powder composition influences the microstructures and properties of the coating. The main components of the coating are WC, Cr,Fe,Ni,Co(W,C), $(Cr,Fe,Ni,Co)_3W_3C$ and the alloy phase of Cr, Fe, Ni and Co. Experimental result shows that the interface between the spray welding layer and matrix has a large amount of WC, and there is a gradient distribution of WC from internal to external in the cross section. The spray welding layer would enhance the hardness and wear resistance of the 42CrMo steel. The hardness of the coatings increased with the mass fraction of the Fe-based amorphous alloy. The spray welding layer has a much higher wear resistance compared with the carbon steel, and Fe-20 exhibits a superior wear resistance when compared to others.

Author Contributions: Data curation, Y.W. and Y.X. (Yi Xu); formal analysis, M.L.; investigation, Y.X. (Yan Xu) and Y.X. (Yi Xu); project administration, Y.X. (Yi Xu) and Z.H. All authors have read and agreed to the published version of the manuscript.

Funding: This research was funded by the Innovation Program (237099000000170004).

Acknowledgments: The authors would like to acknowledge the financial supports of the Innovation Program (237099000000170004).

Conflicts of Interest: The authors declare no conflict of interest.

References

1. Singh, G.; Kumar, S.; Kumar, R. Comparative study of hot corrosion behavior of thermal sprayed alumina and titanium oxide reinforced alumina coatings on boiler steel. *Mater. Res. Express* **2020**, *7*, 26527. [CrossRef]
2. Dong, T.S.; Zheng, X.D.; Li, Y.L.; Li, G.L.; Zhou, X.K.; Wang, H.D. Microstructure and Wear Resistance of FeCrBSi Plasma-Sprayed Coating Remelted by Gas Tungsten Arc Welding Process. *J. Mater. Eng. Perform.* **2018**, *27*, 4069–4076.
3. Jiang, Q.; Tian, Y.; Shu, F.; Zhao, H.; Sun, Y.; He, W.; Xu, B. Strengthening mechanism and properties of Co–WC composite coatings deposited by plasma-transferred arc welding. *Micro Nano Lett.* **2019**, *14*, 717–720. [CrossRef]
4. Zhang, Y.; Chi, Q.; Chang, L.; Dong, Y.; Cai, P.; Pan, Y.; Gong, M.; Huang, J.; Li, J.; He, A.; et al. Novel Fe-based amorphous compound powder cores with enhanced DC bias performance by adding FeCo alloy powder. *J. Magn. Magn. Mater.* **2020**, *507*, 166840. [CrossRef]
5. Leal, E.; Gomes, U.; Alves, S.; Costa, F. The influence of powder preparation condition on densification and microstructural properties of WC-Co- Al2O3 cermets. *Int. J. Refract. Met. Hard Mater.* **2020**, *92*, 105275. [CrossRef]
6. Hu, L.; Huang, J.; Liu, C.; Liu, X.; Hou, D.; Xu, C.; Zhao, Y. Effects of coupling between the laser plasma and two arcs on metal transfer in CO2 laser double-wire MIG hybrid welding. *Opt. Laser Technol.* **2018**, *105*, 152–161. [CrossRef]
7. Kawase, M.; Ido, A.; Morinaga, M. Development of $SiO_2/TiO_2/Al_2O_3$-based/TiO_2 coating for preventing sulfide corrosion in thermal power plant boilers. *Appl. Therm. Eng.* **2019**, *153*, 242–249. [CrossRef]
8. Zhang, S.; Sun, J.; Zhu, M.; Zhang, L.; Nie, P.; Li, Z. Effects of shielding gases on process stability of 10CrNi3MoV steel in hybrid laser-arc welding. *J. Mater. Process. Technol.* **2019**, *270*, 37–46. [CrossRef]
9. Wang, Q.; Zhou, Y.; Zhou, J.; Zhang, G. Microstructure and properties of PTA sprayed 310/WC composite coating. *Mater. Res. Express* **2019**, *6*, 66561. [CrossRef]
10. Liang, X.; Liu, Z.; Wang, B. Multi-pattern failure modes and wear mechanisms of WC-Co tools in dry turning Ti–6Al–4V. *Ceram. Int.* **2020**, *46*, 24512–24525. [CrossRef]
11. Mironchuk, B.; Abrosimova, G.E.; Bozhko, S.; Drozdenko, A.; Postnova, E.; Aronin, A. Phase transformation and surface morphology of amorphous alloys after high pressure torsion. *Mater. Lett.* **2020**, *273*, 127941. [CrossRef]
12. Ma, D.-D.; Xue, Y.-P.; Gao, J.; Ma, Y.; Yu, S.-W.; Wang, Y.-S.; Xue, C.; Hei, H.-J.; Tang, B. Effect of Ta diffusion layer on the adhesion properties of diamond coating on WC-Co substrate. *Appl. Surf. Sci.* **2020**, *527*, 146727. [CrossRef]
13. Methong, T.; Shigeta, M.; Tanaka, M.; Ikeda, R.; Matsushita, M.; Poopat, B. Visualization of gas metal arc welding on globular to spray transition current. *Sci. Technol. Weld. Join.* **2017**, *23*, 87–94. [CrossRef]
14. Staia, M.; Mejias, A.; La Barbera, J.G.; Puchi-Cabrera, E.; Villalobos-Gutierrez, C.; Santana, Y.Y.; Montagne, A.; Iost, A.; Rodriguez, M.A. Mechanical properties of WC/Co-CNT HVOF sprayed coatings. *Surf. Eng.* **2018**, *36*, 1156–1164. [CrossRef]
15. Yan, W.; Qin, H.; Qiang, X.; Zhong, X. Microstructures and Wear Behavior of Ni-based Spray-welding Coating on Pure Titanium TA1 Substrate. *Rare Metal. Mat. Eng.* **2018**, *47*, 910–914.
16. Maslarevic, A.; Bakic, G.M.; Djukic, M.B.; Rajicic, B.; Maksimovic, V.; Pavkov, V. Microstructure and Wear Behavior of MMC Coatings Deposited by Plasma Transferred Arc Welding and Thermal Flame Spraying Processes. *Trans. Indian Inst. Met.* **2020**, *73*, 259–271. [CrossRef]
17. Kendzia, B.; Koppisch, D.; Van Gelder, R.; Gabriel, S.; Zschiesche, W.; Behrens, T.; Brüning, T.; Pesch, B. Modelling of exposure to respirable and inhalable welding fumes at German workplaces. *J. Occup. Environ. Hyg.* **2019**, *16*, 400–409. [CrossRef]
18. Yu, J.; Wang, B.; Zhang, H.; Wang, Q.; Wei, L.; Chen, P.; He, P.; Feng, J. Characteristics of Magnetic Field Assisting Plasma GMAW-P The effect of the magnetic field intensity on droplet transition in plasma-GMAW-P hybrid welding was studied. *Weld J.* **2020**, *99*, S25–S38. [CrossRef]
19. Bayar, I.; Ulutan, M. Surface modification of atmospheric plasma sprayed cermet coatings by plasma transferred arc method. *Trans. IMF* **2019**, *97*, 298–304. [CrossRef]
20. Baiamonte, L.; Tului, M.; Bartuli, C.; Marini, D.; Marino, A.L.; Menchetti, F.; Pileggi, R.; Pulci, G.; Marra, F. Tribological and high-temperature mechanical characterization of cold sprayed and PTA-deposited Stellite coatings. *Surf. Coatings Technol.* **2019**, *371*, 322–332. [CrossRef]
21. Huang, E.-W.; Hung, G.-Y.; Lee, S.Y.; Jain, J.; Chang, K.-P.; Chou, J.J.; Yang, W.-C.; Liaw, P.K. Mechanical and Magnetic Properties of the High-Entropy Alloys for Combinatorial Approaches. *Crystals* **2020**, *10*, 200. [CrossRef]
22. Zhao, J.; Gao, Q.; Wang, H.; Shu, F.; Zhao, H.; He, W.; Yu, Z. Microstructure and mechanical properties of Co-based alloy coatings fabricated by laser cladding and plasma arc spray welding. *J. Alloys Compd.* **2019**, *785*, 846–854. [CrossRef]
23. Dong, S.; Jiang, F.; Xu, B.; Chen, S. Influence of Polarity Arrangement of Inter-Wire Arc on Droplet Transfer in Cross-Coupling Arc Welding. *Materials* **2019**, *12*, 3985. [CrossRef] [PubMed]
24. Zhao, Y.; Chung, H. Numerical simulation of the transition of metal transfer from globular to spray mode in gas metal arc welding using phase field method. *J. Mater. Process. Technol.* **2018**, *251*, 251–261. [CrossRef]
25. Zhu, R.; Gao, W. Wear-resistance Performance of Spray-welding Coating by Plasma Weld-surfacing. *J. Wuhan Univ. Technol. Sci. Ed.* **2018**, *33*, 414–418. [CrossRef]
26. Shi, B.; Huang, S.; Zhu, P.; Xu, C.; Guo, P.; Fu, Y. In-situ TiN reinforced composite coatings prepared by plasma spray welding on Ti6Al4V. *Mater. Lett.* **2020**, *276*, 128093. [CrossRef]
27. Chu, Z.; Deng, W.; Zheng, X.; Zhou, Y.; Zhang, C.; Xu, J.; Gao, L. Corrosion Mechanism of Plasma-Sprayed Fe-Based Amorphous Coatings with High Corrosion Resistance. *J. Therm. Spray Technol.* **2020**, *29*, 1111–1118. [CrossRef]

28. Xu, L.P.; Song, J.B.; Deng, C.G.; Liu, M.; Zhou, K.S. Microstructure and Properties of WC-based Coating Reinforced by Fe-based Amorphous Alloys. *Rare Metal Mat. Eng.* **2020**, *49*, 1546–1552.
29. Kim, W.R.; Heo, S.; Kim, J.-H.; Park, I.-W.; Chung, W. Multi-Functional Cr-Al-Ti-Si-N Nanocomposite Films Deposited on WC-Co Substrate for Cutting Tools. *J. Nanosci. Nanotechnol.* **2020**, *20*, 4390–4393. [CrossRef]
30. Zhang, L.; Yue, F.; Li, S.F.; Yang, Y.F. Utilizing the autocatalysis of Co to prepare low-cost WC-Co powder for high-performance atmospheric plasma spraying. *J. Am. Ceram. Soc.* **2020**, *103*, 6690–6699. [CrossRef]
31. Hu, M.; Tang, J.; Chen, X.-G.; Ye, N.; Zhao, X.-Y.; Xu, M.-M. Microstructure and properties of WC-12Co composite coatings prepared by laser cladding. *Trans. Nonferrous Met. Soc. China* **2020**, *30*, 1017–1030. [CrossRef]
32. Mostajeran, A.; Shoja-Razavi, R.; Hadi, M.; Erfanmanesh, M.; Barekat, M.; Firouzabadi, M.S. Evaluation of the mechanical properties of WC-FeAl composite coating fabricated by laser cladding method. *Int. J. Refract. Met. Hard Mater.* **2020**, *88*, 105199. [CrossRef]
33. Huang, F.; Xu, H.; Liu, W.; Zheng, S. Microscopic characteristics and properties of titaniferous compound reinforced nickel-based wear-resisting layer via in situ precipitation of plasma spray welding. *Ceram. Int.* **2018**, *44*, 7088–7097. [CrossRef]
34. Boukantar, A.-R.; Djerdjare, B.; Guiberteau, F.; Ortiz, A.L. Spark plasma sinterability and dry sliding-wear resistance of WC densified with Co, Co+Ni, and Co+Ni+Cr. *Int. J. Refract. Met. Hard Mater.* **2020**, *92*, 105280. [CrossRef]
35. Hua, N.; Zhang, X.; Liao, Z.; Hong, X.; Guo, Q.; Huang, Y.; Ye, X.; Chen, W.; Zhang, T.; Jin, X.; et al. Dry wear behavior and mechanism of a Fe-based bulk metallic glass: Description by Hertzian contact calculation and finite-element method simulation. *J. Non-Cryst. Solids* **2020**, *543*, 120065. [CrossRef]
36. Itagaki, H.; Yachi, T.; Ogiso, H.; Sato, H.; Yamashita, Y.; Yasuoka, J.; Funada, Y. DC Arc Plasma Treatment for Defect Reduction in WC-Co Granulated Powder. *Metals* **2020**, *10*, 975. [CrossRef]
37. Fu, W.; Chen, Q.-Y.; Yang, C.; Yi, D.-L.; Yao, H.-L.; Wang, H.-T.; Ji, G.-C.; Wang, F. Microstructure and properties of high velocity oxygen fuel sprayed (WC-Co)-Ni coatings. *Ceram. Int.* **2020**, *46*, 14940–14948. [CrossRef]
38. Li, C.; Wang, H.; Ding, J.; Wang, S.; Li, J.; Kou, S. Effects of heat treatment on HVOF-sprayed Fe-based amorphous coatings. *Surf. Eng.* **2020**. [CrossRef]
39. Derakhshandeh, M.; Eshraghi, M.; Jam, A.; Rajaei, H.; Fazili, A. Comparative studies on corrosion and tribological performance of multilayer hard coatings grown on WC-Co hardmetals. *Int. J. Refract. Met. Hard Mater.* **2020**, *92*, 105339. [CrossRef]
40. Ning, W.; Zhai, H.; Xiao, R.; He, D.; Liang, G.; Wu, Y.; Li, W.; Li, X. The Corrosion Resistance Mechanism of Fe-Based Amorphous Coatings Synthesised by Detonation Gun Spraying. *J. Mater. Eng. Perform.* **2020**, *29*, 3921–3929. [CrossRef]
41. Ogino, Y.; Hirata, Y.; Asai, S. Discussion of the Effect of Shielding Gas and Conductivity of Vapor Core on Metal Transfer Phenomena in Gas Metal Arc Welding by Numerical Simulation. *Plasma Chem. Plasma Process.* **2020**, *40*, 1109–1126. [CrossRef]
42. Shi, B.; Huang, S.; Zhu, P.; Xu, C.; Zhang, T. Microstructure and Wear Behavior of In-Situ NbC Reinforced Composite Coatings. *Materials* **2020**, *13*, 3459. [CrossRef] [PubMed]
43. Stummer, M.; Stütz, M.; Aumayr, A.; Enzinger, N. Electron beam welding of copper using plasma spraying for filler metal deposition. *Weld. World* **2018**, *62*, 1341–1350. [CrossRef]

 processes

MDPI

Article

Time of Flight Size Control of Carbon Nanoparticles Using Ar+CH_4 Multi-Hollow Discharge Plasma Chemical Vapor Deposition Method

Sung Hwa Hwang [1], Kazunori Koga [1,2,*], Yuan Hao [1], Pankaj Attri [3], Takamasa Okumura [1], Kunihiro Kamataki [1], Naho Itagaki [1], Masaharu Shiratani [1], Jun-Seok Oh [4], Susumu Takabayashi [5] and Tatsuyuki Nakatani [6]

1 Department of Electronics, Kyushu University, Fukuoka 819-0395, Japan; sh.hwang@plasma.ed.kyushu-u.ac.jp (S.H.H.); y.hao@plasma.ed.kyushu-u.ac.jp (Y.H.); t.okumura@plasma.ed.kyushu-u.ac.jp (T.O.); kamataki@plasma.ed.kyushu-u.ac.jp (K.K.); itagaki@ed.kyushu-u.ac.jp (N.I.); siratani@ed.kyushu-u.ac.jp (M.S.)
2 Center for Novel Science Initiatives, National Institutes of Natural Science, Tokyo 105-0001, Japan
3 Center of Plasma Nano-Interface Engineering, Kyushu University, Fukuoka 819-0395, Japan; attri.pankaj.486@m.kyushu-u.ac.jp
4 Graduate School of Engineering, Osaka City University, Osaka 558-8585, Japan; jsoh@osaka-cu.ac.jp
5 National Institute of Technology, Ariake College, Fukuoka 836-8585, Japan; stak@ariake-nct.ac.jp
6 Institute of Frontier Science and Technology, Okayama University of Science, Okayama 700-0005, Japan; nakatani@bme.ous.ac.jp
* Correspondence: koga@ed.kyushu-u.ac.jp

Abstract: As the application of nanotechnology increases continuously, the need for controlled size nanoparticles also increases. Therefore, in this work, we discussed the growth mechanism of carbon nanoparticles generated in Ar+CH_4 multi-hollow discharge plasmas. Using the plasmas, we succeeded in continuous generation of hydrogenated amorphous carbon nanoparticles with controlled size (25–220 nm) by the gas flow. Among the nanoparticle growth processes in plasmas, we confirmed the deposition of carbon-related radicals was the dominant process for the method. The size of nanoparticles was proportional to the gas residence time in holes of the discharge electrode. The radical deposition developed the nucleated nanoparticles during their transport in discharges, and the time of flight in discharges controlled the size of nanoparticles.

Keywords: plasma chemical vapor deposition; carbon nanoparticle; coagulation; optical emission spectroscopy

Citation: Hwang, S.H.; Koga, K.; Hao, Y.; Attri, P.; Okumura, T.; Kamataki, K.; Itagaki, N.; Shiratani, M.; Oh, J.; Takabayashi, S.; Nakatani, T.; et al. Time of Flight Size Control of Carbon Nanoparticles Using Ar+CH_4 Multi-Hollow Discharge Plasma Chemical Vapor Deposition Method. *Processes* **2021**, *9*, 2. https://dx.doi.org/10.3390/pr9010002

Received: 13 November 2020
Accepted: 16 December 2020
Published: 22 December 2020

1. Introduction

Carbon nanoparticles (CNPs) have attracted tremendous attention for their various applications, such as electrical conductivity improvement of polymer, lubrication applications, cancer cell treatments, bioimaging diagnostics [1–3]. Therefore, it is essential to develop a simple method to control the size and structure of CNPs [4–7]. The solution process is a conventional method of producing CNPs, but this method has limitations like impurity, unexpected agglomeration, and low throughput due to the multistage process [8–11].

The plasma process plays a promising role because it is a dry process using low pressure resulting from reducing impurity and avoid agglomeration or coagulation due to the charge of CNPs. However, the traditional plasma process has a problem regarding throughput due to pulsed discharges for size control [12–14]. Traditional plasma process has the discharge off period to wait for pumping out the particles from the gas phase, resulting in lower throughput.

To date, we have successfully synthesized Si NPs and CNPs by using multi-hollow discharge plasma chemical vapor deposition (MHDPCVD), which can be produced continuously by employing fast gas flow [15–23]. In this method, the gas flow direction is uniform in the plasma region, and NPs are nucleated and grown in plasma. The nucleated NPs were transported toward the outside of the plasmas by viscous gas force. That results in stopped growth outside of the plasma, which helped in the continuous production of size-controlled NPs.

NPs synthesis by the MHDPCVD method undergoes parametric tests such as dependence on gas pressure, gas flow rate, and gas composition, which are the external parameters [15–24]. Using the MHDPCVD, crystalline Si nanoparticles of 2 nm in size with 0.5 nm in size dispersion were produced for nanocrystalline amorphous silicon films for the third generation solar cells [15–21]. We employed two MHDPCVD sources to produce size-controlled Si nanoparticles and to cover nitrogen on the particles. The surface-modified nanoparticles showed multi-exciton generation, which is necessary to increase solar cells' efficiency [18]. We recently used MHDPCVD to produce carbon nanoparticles [23–25] and confirmed that pressure played an important role in size control [23]. These studies revealed the essential parameters for the size control, while the growth mechanism was unclear. Hence, in this study, we measured the CNPs size dependence on the gas flow rate (*FR*) and discussed the growth mechanism of nanoparticles produced by the MHDPCVD method.

2. Materials and Methods

Figure 1 illustrates a schematic diagram of the MHDPCVD reactor [23,24]. Powered and grounded electrodes have 8 holes of 5 mm diameter. The powered electrode of 5 mm in thickness was a sandwich between two grounded electrodes of 1 mm in thickness. The gap between the powered and grounded electrode was 2 mm, and the total length of a hole was 11 mm. Ar and CH_4 gases were introduced from the chamber's left side, passed through the holes, and later evacuated by the pump system. The *FR* ratio of Ar and CH_4 was 6:1. The total *FR* was controlled in a range of 10–120 sccm. During this process, gas pressure was kept at 266 Pa. The substrate holder was set at 100 mm apart from the electrode in the downstream region, and it was grounded. The powered electrode was connected to a 60 MHz radio frequency (rf) power supply through an impedance matching box. The discharge power and discharge period were 40 W and 90 min, respectively, and corresponding discharge and self-bias voltages were 230 and 80 V, respectively.

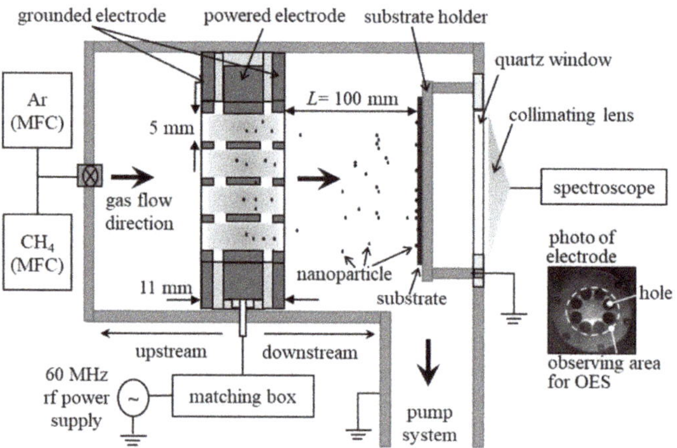

Figure 1. Multi-hollow discharge plasma.

The CNPs generated in the discharges were collected by using mesh grids for a transmission electron microscope (TEM) and Si substrates for a Raman spectroscopy. The size and structure of CNPs were measured with TEM (JEOL, JEM-2010) and Raman spectroscope (Jasco, NRS3000; $\lambda = 532$ nm), respectively. Optical emission from plasmas discharges (all eight holes) was monitored by spectroscope (Ocean Optics, USB2000+) equipped with a collimating lens.

3. Results and Discussion

Figure 2a–d show the TEM images of CNPs as an FR parameter. With increasing *FR*, the mean size of CNPs decreased from approximately 220 nm at *FR* = 10 sccm to 25 nm at *FR* = 120 sccm. For *FR* = 10 sccm (Figure 2a) and *FR* = 20 sccm (Figure 2b), the CNPs deposit sparsely, while they deposit densely for *FR* = 50 sccm (Figure 2c) and *FR* = 120 sccm (Figure 2d). Additionally, the number of deposited CNPs increased with increasing *FR*. This shows the flux of CNPs increases with the increase in *FR*. Above 50 sccm, the deposited CNPs were stacked, then the absolute number of the deposited CNPs were unclear. Thus, we have evaluated the probability distribution of the CNPs for each *FR*.

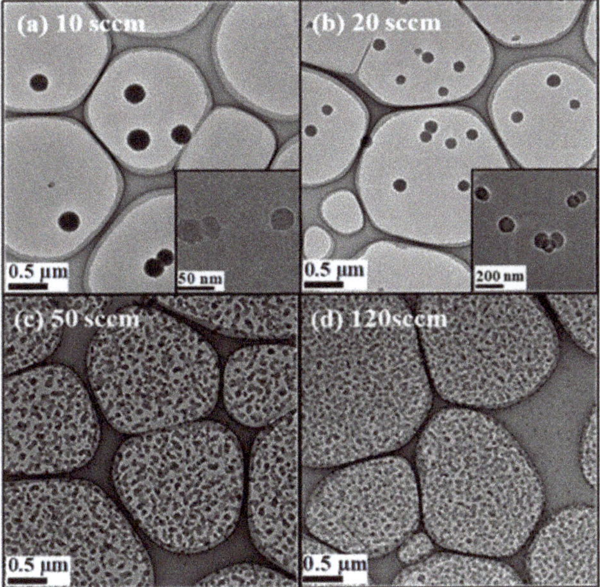

Figure 2. TEM images of carbon nanoparticles produced for (**a**) *FR* = 10 sccm, (**b**) *FR* = 20 sccm, (**c**) *FR* = 50 sccm, and (**d**) *FR* = 120 sccm. Insets in (**a**) and (**b**) show their high magnification TEM images.

Figure 3 shows the size distribution of the deposited CNPs obtained from TEM images where d_p is the CNP size (diameter). The deposited nanoparticles were stacked for *FR* above 50 sccm. Thus, we estimated the probability of CNPs deposited on the mesh grid. Two group sizes were produced for *FR* = 10 sccm; (1) smaller group size has a size range between 20 and 90 nm, and (2) larger group size has a range between 170 and 250 nm. For *FR* = 20 sccm, two peaks at 60 and 150 nm were detected, but these peaks overlap and form one size group with a wide range between 30–200 nm. At the same time, one group size was obtained for *FR* above 50 sccm. Therefore, as the *FR* increases from 50 to 120 sccm, the peak size gradually shifts toward a smaller size from 45 to 20 nm, respectively. The size dispersion became narrower for higher *FR* from 50 and 120 sccm.

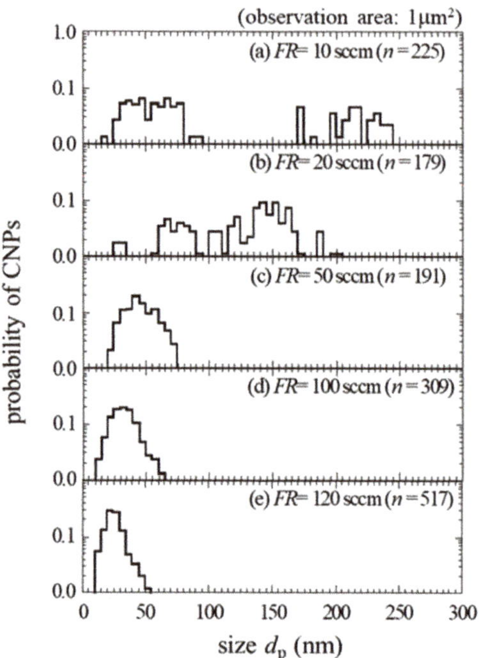

Figure 3. The size distribution of CNPs as a parameter of *FR* where n is the number of measured CNPs.

From the size distribution in Figure 3, we plotted a dependence of d_p on *FR*, as shown in Figure 4. At *FR* below 20 sccm, the larger-sized nanoparticles seem to be separated from the smaller size group and grow in a monodisperse way. Similar growth behavior was observed for Si nanoparticles in silane plasmas in the earlier study [26]. Considering the larger size of CNPs at *FR* = 10 sccm, the d_p decreases monotonically with increasing *FR*.

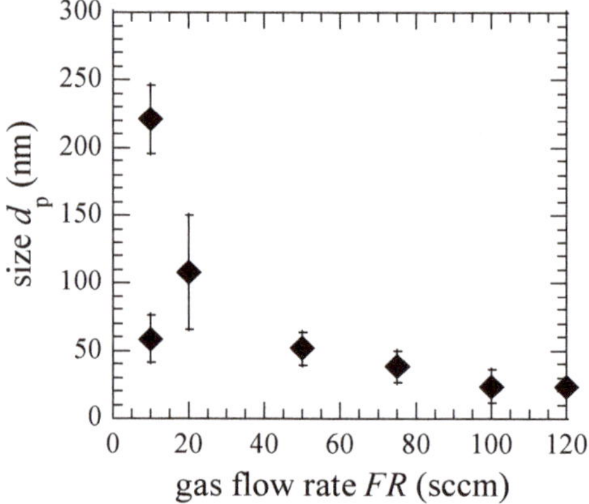

Figure 4. Dependence of d_p on *FR*. Error bar shows the standard deviation.

To obtain the structure of the CNPs, we have measured the XRD and Raman spectra of CNPs deposited at *FR* = 50 sccm. Figures 5 and 6 show the XRD and Raman spectra, respectively. A broad peak in the XRD spectrum appears around 2θ = 20° and corresponds to the hydrogenated amorphous carbon (a-C:H) [27]. Figure 6 shows a Raman spectrum of nanoparticles deposited on the Si substrates at 50 sccm *FR*. Raman spectra clearly show the separated D (1350 cm^{-1}), and G (1580 cm^{-1}) bands. The area intensity ratio of D/G band was around 1.8; this indicates the structure of the CNPs were polymer like a-C:H [25,28–30]. Similar spectra were also observed at other *FR*s.

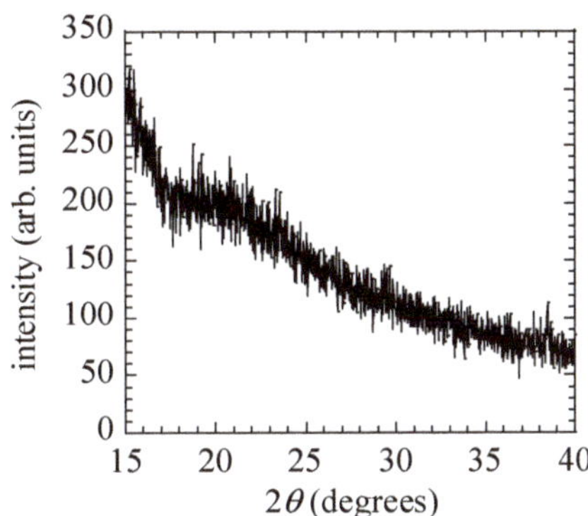

Figure 5. XRD spectrum of CNPs for *FR* = 50 sccm.

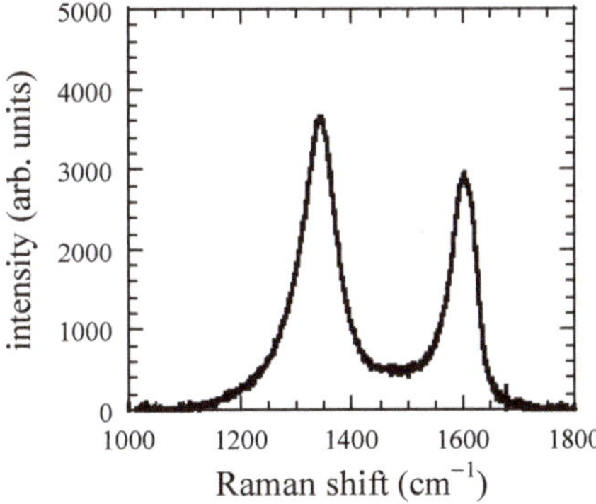

Figure 6. Raman spectrum of CNPs for *FR* = 50 sccm.

For the nanoparticle growth in the conventional CCP, the discharge duration is an essential factor. The size of nanoparticles increasing with an increase in CCP discharge duration [31]. The discharge duration was related to the period, which is the sum of the nucleation time and subsequent nucleated nanoparticles' growth time. Continuous discharges sustained in holes and a low-density plasma penetrated the holes due to high working pressure of 266 Pa. The generated CNPs were transported inside the holes by the gas flow. The growth time of CNPs in the plasmas correlates with the gas residence time in holes. The gas residence time τ_{res} of holes corresponds to discharge duration in the conventional CCP. In this study, gas residence time was calculated from *FR*. For the CNP, growth involves two growth processes like the coagulation of CNPs during transport toward substrates and radical deposition on CNPs.

For the coagulation, CNPs are grown by the collision between two CNPs, as the volume of CNPs after the collision is the sum of two CNPs volume (before the collision). The size d_{p1} and number density n_{p1} of CNPs after the collision are expressed by $d_{p1} = 2^{\frac{1}{3}} d_{p0}$ and $n_{p1} = n_{p0} - 1$, respectively, where d_{p0} and n_{p0} are the size and density of CNPs before the collision. To figure out the effects of the coagulation of CNPs, we examined the CNPs deposition at three positions in the transport region. Figure 7 shows the dependence of the size and surface density of deposited CNPs on the position L far from the electrode. For L = 100, 120, and 140 mm, the size is irrelevant to the position. The area density monotonically decreases with increasing L.

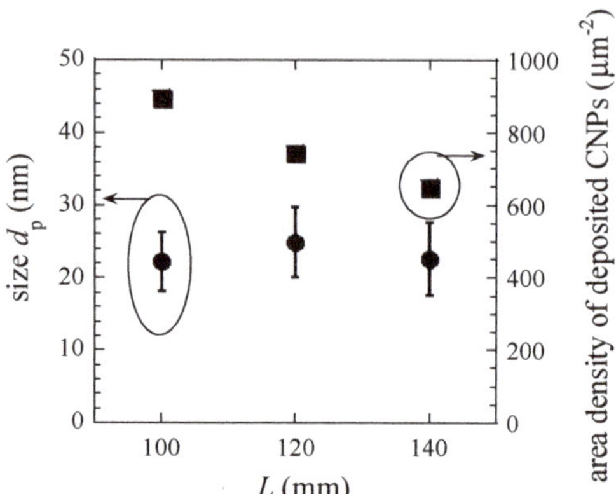

Figure 7. Dependence of the size and surface density of deposited CNPs on the position L far from the electrode for FR = 100 sccm. Error bar shows the standard deviation.

Previously, it was reported that the flux of CNPs proportionally increases with increasing the solid angle in the multi-hollow discharge plasma CVD method [24]. The solid angle is the main factor of decreased area density, see Figure 7. The results of the size and the area density indicate that the coagulation of CNPs was negligible. For the radical deposition, the growth rate G_r of CNP expressed as equation 1

$$G_r = \frac{dd_p}{dt} = 2DR_r, \tag{1}$$

where d_p is the size (diameter) of CNPs and DR_r is the deposition rate of radicals on CNPs. If we assume the sticking probability of radicals on CNPs is unity and carbon atoms are responsible for the mass of CNPs, the G_r is given by

$$G_r = \frac{dd_p}{dt} = \frac{2}{\rho} m_C n_r v_{thr}, \tag{2}$$

where ρ is the mass density of CNPs, m_C the mass of a carbon atom (2.00×10^{-26} kg), n_r the number density of the radicals in plasmas, and v_{thr} the thermal velocity of the radicals. The size and density of CNPs affect the radical density. The loss of radicals to the chamber wall is dominant if their size and density are low, while the loss to CNPs is prevalent if their size and density are high. The loss mode is determined by the coupling parameter Γ of CNPs in plasmas [32], given by the following equation.

$$\Gamma = \frac{1}{6} d_p^2 n_p^{\frac{5}{3}} D_w^3, \tag{3}$$

where D_w is the characteristic length of the reactor. For the $\Gamma >> 1$, the coupling among CNPs through radicals is strong, results in the deceased radical density with the time after the nucleation of CNPs. If the coupling is weak, the wall loss of radicals is dominant, resulting in no radical density change with the time. Further, to detect the Γ value, the n_p was deduced from the result in Figure 7.

Figure 7 shows the number of deposited CNPs per μm^2 during the deposition time of 60 min. The flux of the CNPs can be calculated if the sticking probability of CNPs is unity. Considering the solid angle, the flux at the end of the holes deduced to be

1.30×10^{11} cm^{-2}s^{-1}. Raman results show that the structure of CNPs was polymer-like carbon, and the mass density of the CNPs was assumed 1.6 g/cm^3. If the temperature of CNPs equal to that of the electrode (433 K), the n_p was 1.20×10^9 cm^{-3}, and d_p was 25 nm, as shown in Figure 6. D_w (D_w = 2.5 mm) assumed as the radius of hole, then Γ was 6.78 at the end of the discharge region. In the discharge region, the size of CNPs was smaller than 25 nm, and Γ value should be less than one. It suggests that the loss of radicals through the wall was predominant, which results in a constant rate of radical loss. To discuss the generation of the radicals, we have measured emission spectra in plasma. We measured two Ar I emission intensities at 425.9 nm $I_{425.9}$ and 750.4 nm $I_{750.4}$ with upper-level excitation energy of 14.7 eV (3p$_1$) and 13.5 eV (2p$_1$), respectively. These emission processes have little effect on quenching and radiation trapping. The upper excitation level has small cross sections for electron-impact excitation from metastable states [33,34]. The *FR* dependence of an emission intensity ratio $I_{425.9}/I_{750.4}$, shown in Figure 8. The ratio indicates the information of the high energy tail of the electron energy distribution, which relates to the radical generation.

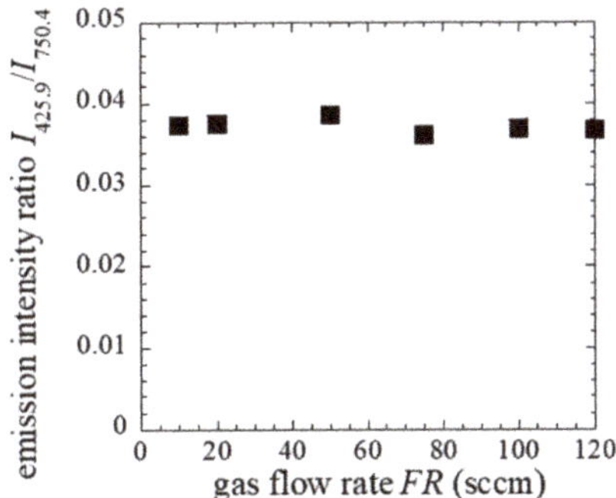

Figure 8. Dependence of $I_{425.9}/I_{750.4}$ on *FR*.

Although the ratio is irrelevant to *FR*, as shown in Figure 8, the discharge voltage for each *FR* condition is almost the same, suggesting the electron density was irrelevant to the *FR*. These results indicate that the generation rate of radicals is unrelated to the *FR*.

In the steady state, the n_r is proportional to the density of CH_4 because the electron density and the loss rate of the radicals can be assumed to be constant based on the above discussion. Integrating Equation (2), the following formula gives the CNP size.

$$d_p = \frac{2}{\rho} k m_C n_{CH4} v_{thr} t, \tag{4}$$

where k is the ratio of generation rate and loss rate of radicals in the steady state, n_{CH4} the density of CH_4, and t the interaction time of CNPs and radicals. The k value is related to the depletion rate of the CH_4 molecules. In the current study, CNPs were nucleated in the discharge generated in the holes of the electrode. They grew in the discharge, transport with the gas flow, and growth was stopped outside the holes. We assumed that the growth of CNPs starts when the CH_4 molecules enter into the holes where plasmas were generated. Thus, d_p was assumed to be equal to zero at t = 0, and the growth of CNPs stops at $t = \tau_{res}$. Figure 8 shows the dependence of d_p on τ_{res}, based on *FR* dependence, together with the results reported earlier [23]. Considering the larger size of CNPs for *FR* = 10 sccm, the size of CNPs linearly increases with increasing the τ_{res}. In this study, n_{CH4} and v_{thr} was 6.36×10^{21} m^{-3} and 8.24×10^2 m/s, respectively. The calculated value using Equation (4) as a parameter of k and the experimental results were well fitted for k = 0.035 (Figure 9). For the conventional CCP, the depletion rate of CH_4 was about 3% for

1.33 Pa pure CH_4 gas and 0.15 W/cm^2 in discharge power density [35]. The depletion rate monotonically increases with CH_4 pressure.

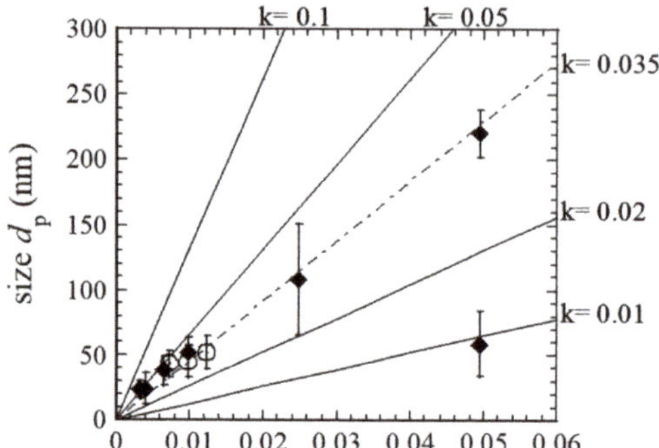

Figure 9. Dependence of d_p on τ_{res}. Open circles indicate the results reported earlier [23]. Lines were obtained by Equation (4). Error bar shows the standard deviation.

For the MHDPCVD, the discharge power density was 6.4 W/cm^2, much higher than the conventional plasma CVD (above-mentioned), and the partial pressure of CH_4 was 38 Pa. The radical loss to CNPs was small but cannot be ignored as it affects the Γ value. Thus, the fitted value of k is reasonable. Based on our results, the CNP in MHDPCVD was grown by the deposition of carbon-related radicals.

4. Conclusions

Through this work, we succeeded in synthesizing the size controlled CNPs using the Ar +CH4 MHDPCVD method continuously. The control range of the mean size was from 25 to 220 nm. We observed that size was proportional to the gas residence time in the discharges maintained in the electrode's holes. We theoretically confirmed that CNPs were grown by the deposition of radicals during the discharges' transport of CNPs, and CNPs move through gas flow in the discharges. The duration of the CNP transport in the discharge corresponds to the gas residence time. Therefore, the CNP size control using the MHDPVCD is a type of time of flight size control.

Author Contributions: Formal analysis, S.H.H. and Y.H.; writing—original draft preparation, S.H.H. and K.K. (Kazunori Koga); writing—review and editing, P.A., T.O., K.K. (Kunihiro Kamataki), N.I., M.S., J.-S.O., S.T. and T.N.; supervision, K.K. (Kazunori Koga), M.S. and T.N.; funding acquisition, K.K. (Kazunori Koga) and M.S. All authors have read and agreed to the published version of the manuscript.

Funding: This study was partly supported by JSPS KAKENHI Grant Number JP20H00142, JP20J13122 and JSPS Core-to-Core Program JPJSCCA2019002.

Institutional Review Board Statement: Not applicable.

Informed Consent Statement: Not applicable.

Data Availability Statements: Data is contained within the article.

Acknowledgments: This study was encouraged by Advanced Characterization Platform of the Nanotechnology Platform Japan sponsored by the Ministry of Education, Culture, Sports, Science and Technology (MEXT), Japan.

Conflicts of Interest: The authors declare no conflict of interest.

References

1. Kumar, S.; Nehra, M.; Kedia, D.; Dilbaghi, N.; Tankeshwar, K.; Kim, K.H. Nanodiamonds: Emerging face of future nanotechnology. *Carbon N. Y.* **2019**, *143*, 678–699. [CrossRef]
2. Li, S.; Pasc, A.; Fierro, V.; Celzard, A. Hollow carbon spheres, synthesis and applications-a review. *J. Mater. Chem. A* **2016**, *4*, 12686–12713. [CrossRef]
3. Bian, X.; Chen, Q.; Zhang, Y.; Sang, L.; Tang, W. Deposition of nano-diamond-like carbon films by an atmospheric pressure plasma gun and diagnostic by optical emission spectrum on the process. *Surf. Coat. Technol.* **2008**, *202*, 5383–5385. [CrossRef]
4. Tian, F.; He, C.N. Fabrication and growth mechanism of carbon nanospheres by chemical vapor deposition. *Mater. Chem. Phys.* **2010**, *123*, 351–355. [CrossRef]
5. Wang, F.; Hong, R. Continuous preparation of structure-controlled carbon nanoparticle via arc plasma and the reinforcement of polymeric composites. *Chem. Eng. J.* **2017**, *328*, 1098–1111. [CrossRef]
6. Estes, C.S.; Gerard, A.Y.; Godward, J.D.; Hayes, S.B.; Liles, S.H.; Shelton, J.L.; Stewart, T.S.; Webster, R.I.; Webster, H.F. Preparation of highly functionalized carbon nanoparticles using a one-step acid dehydration of glycerol. *Carbon N. Y.* **2019**, *142*, 547–557. [CrossRef]
7. Li, Z.; Wang, L.; Li, Y.; Feng, Y.; Feng, W. Carbon-based functional nanomaterials: Preparation, properties and applications. *Compos. Sci. Technol.* **2019**, *179*, 10–40. [CrossRef]
8. Kang, J.; Li, O.L.; Saito, N. Synthesis of structure-controlled carbon nano spheres by solution plasma process. *Carbon N. Y.* **2013**, *60*, 292–298. [CrossRef]
9. Miao, J.Y.; Hwang, D.W.; Narasimhulu, K.V.; Lin, P.I.; Chen, Y.T.; Lin, S.H.; Hwang, L.P. Synthesis and properties of carbon nanospheres grown by CVD using Kaolin supported transition metal catalysts. *Carbon N. Y.* **2004**, *42*, 813–822. [CrossRef]
10. Pottathara, Y.B.; Grohens, Y.; Kokol, V.; Kalarikkal, N.; Thomas, S. *Synthesis and Processing of Emerging Two-Dimensional Nanomaterials*; Elsevier Inc.: Amsterdam, The Netherlands, 2019; ISBN 9780128157510.
11. Kumar, M.; Ando, Y. Chemical Vapor Deposition of Carbon Nanotubes: A Review on Growth Mechanism and Mass Production. *J. Nanosci. Nanotechnol.* **2010**, *10*, 3739–3758. [CrossRef]
12. Koga, K.; Matsuoka, Y.; Tanaka, K.; Shiratani, M.; Watanabe, Y. In situ observation of nucleation and subsequent growth of clusters in silane radio frequency discharges. *Appl. Phys. Lett.* **2000**, *77*, 196–198. [CrossRef]
13. Nunomura, S.; Shiratani, M.; Koga, K.; Kondo, M.; Watanabe, Y. Nanoparticle coagulation in fractionally charged and charge fluctuating dusty plasmas. *Phys. Plasmas* **2008**, *15*, 080703. [CrossRef]
14. Kim, K.; Park, J.H.; Doo, S.G.; Nam, J.D.; Kim, T. Generation of size and structure controlled Si nanoparticles using pulse plasma for energy devices. *Thin Solid Films* **2009**, *517*, 4184–4187. [CrossRef]
15. Nakamura, W.M.; Shimokawa, D.; Miyahara, H.; Koga, K.; Shiratani, M. Spatial Profile of Deposition Rate of a-Si:H Films in Multi-Hollow Discharge Plasma Chemical Vapor Deposition. *Trans. Mater. Res. Soc. Jpn.* **2007**, *32*, 469–472. [CrossRef]
16. Shiratani, M.; Koga, K.; Iwashita, S.; Uchida, G.; Itagaki, N.; Kamataki, K. Nano-factories in plasma: Present status and outlook. *J. Phys. D Appl. Phys.* **2011**, *44*, 1–8. [CrossRef]
17. Koga, K.; Nakahara, K.; Kim, Y.-W.; Kawashima, Y.; Matsunaga, T.; Sato, M.; Yamashita, D.; Matsuzaki, H.; Uchida, G.; Kamataki, K.; et al. Deposition of cluster-free P-doped a-Si:H films using SiH_4+PH_3 multi-hollow discharge plasma CVD. *Phys. Status Solidi* **2011**, *8*, 3013–3016. [CrossRef]
18. Uchida, G.; Yamamoto, K.; Kawashima, Y.; Sato, M.; Nakahara, K.; Kamataki, K.; Itagaki, N.; Koga, K.; Kondo, M.; Shiratani, M. Surface nitridation of silicon nano-particles using double multi-hollow discharge plasma CVD. *Phys. Status Solidi Curr. Top. Solid State Phys.* **2011**, *8*, 3017–3020. [CrossRef]
19. Koga, K.; Matsunaga, T.; Kim, Y.; Nakahara, K.; Yamashita, D.; Matsuzaki, H.; Kamataki, K.; Uchida, G.; Itagaki, N.; Shiratani, M. Combinatorial deposition of microcrystalline silicon films using multihollow discharge plasma chemical vapor deposition. *Jpn. J. Appl. Phys.* **2012**, *51*, 1–4. [CrossRef]
20. Kim, Y.; Matsunaga, T.; Nakahara, K.; Seo, H.; Kamataki, K.; Uchida, G.; Itagaki, N.; Koga, K.; Shiratani, M. Effects of nanoparticle incorporation on properties of microcrystalline films deposited using multi-hollow discharge plasma CVD. *Surf. Coat. Technol.* **2013**, *228*, S550–S553. [CrossRef]
21. Seo, H.; Ichida, D.; Hashimoto, S.; Itagaki, N.; Koga, K.; Shiratani, M.; Nam, S.H.; Boo, J.H. Improvement of charge transportation in Si quantum dot-sensitized solar cells using vanadium doped TiO_2. *J. Nanosci. Nanotechnol.* **2016**, *16*, 4875–4879. [CrossRef]
22. Toko, S.; Kanemitsu, Y.; Seo, H.; Itagaki, N.; Koga, K.; Shiratani, M. Optical bandgap energy of Si nanoparticle composite films deposited by a multi-hollow discharge plasma chemical vapor deposition method. *J. Nanosci. Nanotechnol.* **2016**, *16*, 10753–10757. [CrossRef]
23. Hwang, S.H.; Kamataki, K.; Itagaki, N.; Koga, K.; Shiratani, M. Effects of gas pressure on the size distribution and structure of carbon nanoparticles using Ar + CH_4 multi-hollow discharged plasma chemical vapor deposition. *Plasma Fusion Res.* **2019**, *14*, 1–5. [CrossRef]
24. Hwang, S.H.; Okumura, T.; Kamataki, K.; Itagaki, N.; Koga, K.; Shiratani, M. Size and Flux of Carbon Nanoparticles Synthesized by Ar+CH_4 Multi-hollow Plasma Chemical Vapor Deposition. *Diam. Relat. Mater.* **2020**, 108050. [CrossRef]
25. Hwang, S.H.; Okumura, T.; Kamataki, K.; Itagaki, N.; Koga, K.; Nakatani, T.; Shiratani, M. Low stress diamond-like carbon films containing carbon nanoparticles fabricated by combining rf sputtering and plasma chemical vapor deposition. *Jpn. J. Appl. Phys.* **2020**, *59*, 100906. [CrossRef]

26. Watanabe, Y. Formation and behaviour of nano/micro-particles in low pressure plasmas. *J. Phys. D Appl. Phys.* **2006**, *39*, R329. [CrossRef]
27. Siddique, A.B.; Pramanick, A.K.; Chatterjee, S.; Ray, M. Amorphous Carbon Dots and their Remarkable Ability to Detect 2,4,6-Trinitrophenol. *Sci. Rep.* **2018**, *8*, 9700. [CrossRef]
28. Ferrari, A.C.; Robertson, J. Raman spectroscopy of amorphous, nanostructured, diamond like carbon, and nanodiamond. *Philos. Trans. R. Soc. Lond. A* **2004**, *362*, 2477–2512. [CrossRef]
29. Merlen, A.; Buijnsters, J.G.; Pardanaud, C. A guide to and review of the use of multiwavelength Raman spectroscopy for characterizing defective aromatic carbon solids: From graphene to amorphous carbons. *Coatings* **2017**, *7*, 153. [CrossRef]
30. Dong, X.; Koga, K.; Yamashita, D.; Seo, H.; Itagaki, N.; Shiratani, M.; Setsuhara, Y.; Sekine, M.; Hori, M. Effects of deposition rate and ion bombardment on properties of a-C:H films deposited by H-assisted plasma CVD method. *Jpn. J. Appl. Phys.* **2016**, *55*. [CrossRef]
31. Nunomura, S.; Kita, M.; Koga, K.; Shiratani, M.; Watanabe, Y. In situ simple method for measuring size and density of nanoparticles in reactive plasmas. *J. Appl. Phys.* **2006**, *99*, 083302. [CrossRef]
32. Shiratani, M.; Koga, K.; Kamataki, K.; Iwashita, S. Theory for correlation between plasma fluctuation and fluctuation of nanoparticle growth in reactive plasmas Theory for correlation between plasma fl uctuation and fl uctuation of nanoparticle growth in reactive plasmas. *Jpn. J. Appl. Phys.* **2014**, *53*, 010201. [CrossRef]
33. Moshkalyov, S.A. Deposition of silicon nitride by low-pressure electron cyclotron resonance plasma enhanced chemical vapor deposition in $N_2/Ar/SiH_4$. *J. Vac. Sci. Technol. B Microelectron. Nanom. Struct.* **1997**, *15*, 2682. [CrossRef]
34. Daltrini, A.M.; Moshkalev, S.A.; Monteiro, M.J.R.; Besseler, E.; Kostryukov, A.; Machida, M. Mode transitions and hysteresis in inductively coupled plasmas. *J. Appl. Phys.* **2007**, *101*, 073309. [CrossRef]
35. Toyoda, H.; Kojima, H.; Sugai, H. Mass spectroscopic investigation of the CH_3 radicals in a methane rf discharge. *Appl. Phys. Lett.* **1989**, *54*, 1507–1509. [CrossRef]

Article

Tiny Cold Atmospheric Plasma Jet for Biomedical Applications

Zhitong Chen *,†, Richard Obenchain † and Richard E. Wirz *

Department of Mechanical and Aerospace Engineering, University of California, Los Angeles, CA 90095, USA; robenchain@g.ucla.edu
* Correspondence: zhitongchen@ucla.edu (Z.C.); wirz@ucla.edu (R.E.W.)
† These authors contributed equally to this work.

Abstract: Conventional plasma jets for biomedical applications tend to have several drawbacks, such as high voltages, high gas delivery, large plasma probe volume, and the formation of discharge within the organ. Therefore, it is challenging to employ these jets inside a living organism's body. Thus, we developed a single-electrode tiny plasma jet and evaluated its use for clinical biomedical applications. We investigated the effect of voltage input and flow rate on the jet length and studied the physical parameters of the plasma jet, including discharge voltage, average gas and subject temperature, and optical emissions via spectroscopy (OES). The interactions between the tiny plasma jet and five subjects (de-ionized (DI) water, metal, cardboard, pork belly, and pork muscle) were studied at distances of 10 mm and 15 mm from the jet nozzle. The results showed that the tiny plasma jet caused no damage or burning of tissues, and the ROS/RNS (reactive oxygen/nitrogen species) intensity increased when the distance was lowered from 15 mm to 10 mm. These initial observations establish the tiny plasma jet device as a potentially useful tool in clinical biomedical applications.

Keywords: cold atmospheric plasma; tiny plasma jet; biomedical applications

Citation: Chen, Z.; Obenchain, R.; Wirz, R.E Tiny Cold Atmospheric Plasma Jet for Biomedical Applications. *Processes* **2021**, *9*, 249. https://doi.org/10.3390/pr9020249

Academic Editor: Luisa A. Neves
Received: 8 January 2021
Accepted: 26 January 2021
Published: 29 January 2021

1. Introduction

Plasma is the fourth state of matter (solid, liquid, gas, and plasma), which is the most energetic and abundant state of matter, comprising over 99% of the universe's matter [1]. What makes the plasma unique is its gaseous combination of electrons, ions, and neutral species in both fundamental and excited states [2]. The properties of plasma change depending on the source and amount of energy supplied, and plasma is divided into thermal and non-thermal plasma according to the Maxwell–Boltzmann thermodynamic equilibrium [3]. Depending on the required application, there are a wide variety of plasmas generated under different conditions. Recently, cold atmospheric plasma devices (CAP) operating at atmospheric pressure and room temperature have exhibited great potential for biomedical applications [4–6]. Laroussi demonstrated that plasma generated at atmospheric pressure is a very effective sterilization agent in 1996 [7]. Isbary et al. proposed a first prospective randomized controlled trial to decrease bacterial load using CAP on chronic wounds in patients [8]. Pan et al. developed a novel method of tooth whitening employing CAP driven by direct current in atmospheric pressure air [9]. Chen et al. demonstrated a synergism between CAP and ICB (immune checkpoint blockade) integrated with microneedles to provide a platform technique for the treatment of cancer and other diseases in a minimally invasive manner [10]. More and more manuscripts have been published about CAP on sterilization/disinfection, wound healing, blood coagulation, oral health, cancer therapy, and other applications [11–17]. The efficiency of CAP for biomedical applications mainly relies on its many components, such as reactive nitrogen species (RNS) and reactive oxygen species (ROS) [18–20].

One of the aims of plasma biomedicine is to utilize CAP inside a living organism's body. The conventional form of CAP is not applicable for such in vivo biomedical applications, especially cancer therapy, due to major drawbacks, including high voltages, plasma jet volume, reactive species delivery, and the discharge formation within the organ [21]. To mitigate these problems, we developed a single-electrode tiny plasma jet device for biomedical applications. We then examined how different subjects, including deionized (DI) water, metal, cardboard, and tissue, affected the tiny plasma jet during jet-subject interaction processes.

2. Experimental Section

The single-electrode tiny plasma jet device (Figure 1) was developed at UCLA. It consisted of a powered needle electrode connected to a high voltage transformer, all contained within a 3D-printed polylactic acid (PLA) housing (TAZ 6 from LulzBot, Fargo, ND, USA). The electrode was powered by a tabletop DC power supply (1901B, B&K Precision, Yorba Linda, CA, USA) at 8, 10, and 12 V. The DC-to-AC converter was built at UCLA, and a Chirk Industry transformer (RU3222, Chirk Industry, Taiwan) was used to step-up the voltage by 25× at a nominal frequency of 30 kHz. The input power of the tiny plasma jet was between 6 and 12 W. The plastic (polycarbonate) nozzle had an inner diameter of 1.5 mm and a wall thickness of 0.5 mm. Helium (He, ultra-high purity) was employed as feeding gas. The discharge voltage for the plasma jet was measured using a high voltage probe and oscilloscope.

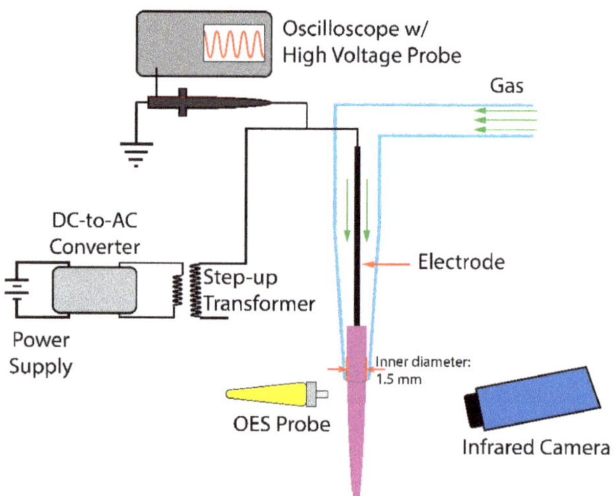

Figure 1. Schematic representation of the single electrode plasma jet device.

A fiber-coupled optical spectrometer (LR1, ASEQ Instruments, Vancouver, BC, Canada), with a range of wavelength 300–1000 nm, was employed to detect CAP-generated ROS and RNS (such as nitric oxide (NO), nitrogen cation (N_2^+), atomic oxygen (O), and hydroxyl radicals (•OH)). The optical probe was placed at a radial distance of 10 mm from the center of the nozzle. Data were collected with an integration time of 600 ms. All subjects but DI water were placed on non-conductive medical-grade plastic for treatment. DI water was placed within a standard plastic 6-cell culture plate.

Thermal measurements of the plasma jet alone and interacting with subjects were made with a long-wavelength infrared camera A655sc, FLIR Systems, Wilsonville, OR, USA) from a distance of approximately 15 cm diagonally above the subject; the relative position of the camera to the jet and subject remained consistent for all tests. Frame sequences for each time point consisted of a multi-second exposure recorded using Research IR 4.40 with individual frames extracted as needed after recording. A linear scale manually configured from 10 °C to 40 °C was selected for all images to allow for a sufficient dynamic range.

Low-magnification optical images of tissue surfaces were made using a BW500 Digital Microscope. Other optical images were taken via digital SLR (D850, Nikon Corporation, Tokyo, Japan) using constant shutter speed, aperture, film speed, and relative position for all images.

3. Results and Discussion

Figure 2a shows the image of the single-electrode tiny plasma jet with a length of approximately 2.5 cm well collimated along the entire length. The thermal effects of the tiny plasma jet on its environment and subjects were measured via thermal camera (Figure 2b). The average gas temperature of the tiny plasma jet is around 25 °C, which is approximately equal to that of the local environment. Figure 2c shows the discharge voltage of the tiny plasma jet at 10 V input voltage and 3.02 L/min He flow rate. The peak-peak discharge voltage is approximately 6.88 kV with a frequency near 20 kHz. The electrode operates with sufficiently high frequency and voltage power to generate a strong electric field, resulting in the formation of the elongated plasma streamer by ionizing the surrounding gas atoms [22]. The magnitude of the current is in milliamperes. Figure 2d shows the optical emission spectra of the tiny plasma jet. The identification of emission lines and bands was performed according to reference [23]. •OH (hydroxyl radical) is present at 309 nm. The He bands are assigned at 588, 668, and 705 nm. The wavelength of 337, 376, and 381 nm could be indicative of the low-intensity N_2 second-positive system ($C^3\Pi_u$-$B^3\Pi_g$). Moreover, their magnitudes are at most a few thousandths of the highest peak N_2^+ (391 nm), while 358 and 428 nm are also N_2^+. The composition of the tiny plasma jet is thus similar to conventional plasma jets [24,25].

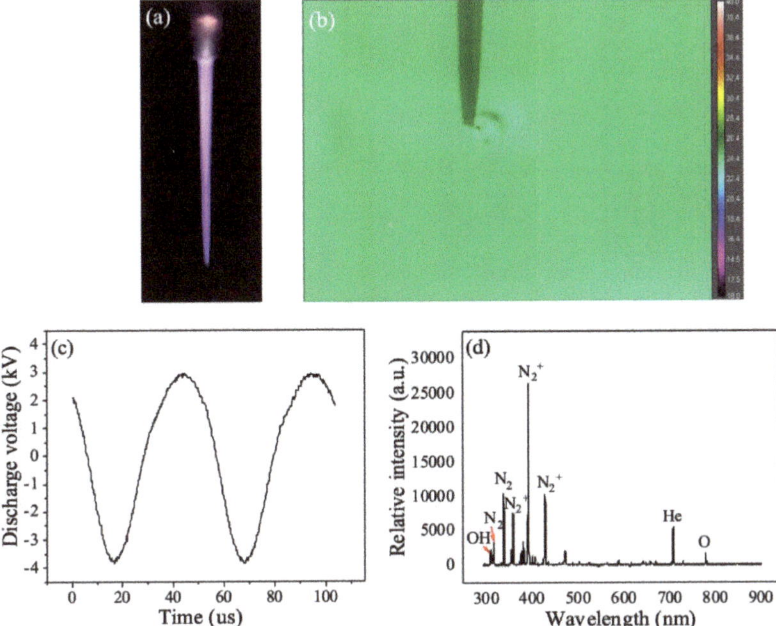

Figure 2. (**a**) Image of the tiny plasma jet and (**b**) thermal measurement of the vicinity of the tiny plasma jet. (**c**,**d**) represent the discharge voltage and optical emission spectrum of tiny plasma jet with input voltage at 10 V and He flow rate at 3.02 L/min, respectively.

Figure 3 shows the effect of input voltage and flow rate on the tiny plasma jet. In Figure 3a, the length of the plasma jet decreases when the flow rate increases at constant input voltage, while the length of the tiny plasma jet increases when the input voltage increases at a constant flow rate. We investigated the longest tiny plasma jet for each input voltage and found that the flow rate for the longest jet is 3.21 L/min for 8 V, 3.02 L/min for 10 V, and 2.51 L/min for 12 V, respectively. Comparing the three longest plasma jets, input voltage plays a major role in affecting the length of the tiny plasma jet. As expected, the flow rate has almost no effect on the peak-peak discharge voltage in Figure 3c; the longest jets for each input have been marked in an ellipse. The authors also pointed out that input voltage plays a significant effect on the length of the plasma plume, among other factors, including flow rate and the diameter of the nozzle exit [26]. The upstream region discharge affects the stability and length of the downstream plasma jet [27]. We used the tiny plasma jet at 10 V input voltage and 3.02 L/min for the following measurements.

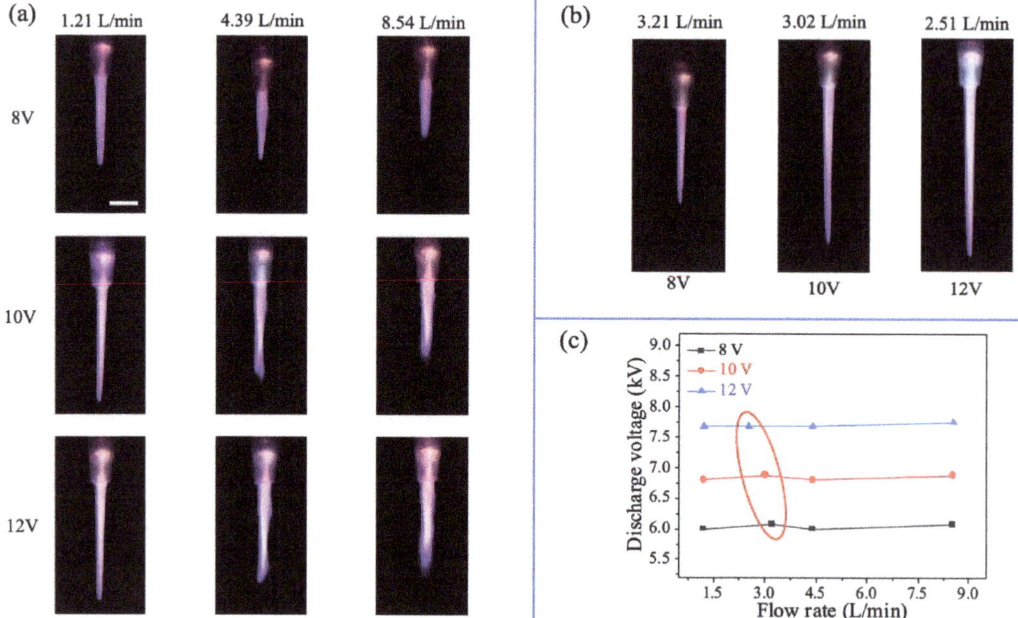

Figure 3. (**a**) The effect of flow rate (1.21 L/min, 4.39 L/min, and 8.54 L/min) and input voltage (8 V, 10 V, and 12 V) on the length of the tiny plasma jet. (**b**) The longest tiny plasma jet for 8 V, 10 V, and 12 V input. (**c**) The peak-peak discharge voltage for the tiny plasma jet at a different He flow rate. Scale bar: 5 mm.

Water is the main component of many living things, exemplified by blood being primarily composed of water [28]. Therefore, it is necessary to study the effect of water on the tiny plasma jet for future biomedical applications. In addition, the conductivity of subjects cannot be controlled during the use of the tiny plasma jet for the treatment of diseases. Here, we selected an aluminum cutout and length of cardboard as conductive and non-conductive subjects, respectively. Figure 4a shows images of the tiny plasma jet interacting with DI water, metal, and cardboard at 10 mm and 15 mm distances. Compared with the tiny plasma alone, the peak-peak discharge voltages of the jet interacting with DI water, metal, and cardboard decrease slightly at the 15 mm distance, while the peak-peak discharge voltage decreases more for each one at 10 mm distance compared with 15 mm. From a 15 mm distance to 10 mm, plasma intensity increases, especially for DI water and metal, becoming visibly brighter. Figure 4b shows the optical emission spectra of the tiny plasma jet interacting with DI water, metal, and cardboard at 10 mm and 15 mm distances. At a 15 mm distance, the ROS/RNS intensity of the tiny plasma jet interacting with DI water, metal, and cardboard only varies slightly when compared with the tiny plasma jet operating without a subject. From 15 mm to 10 mm, the ROS/RNS intensity increases slightly for jet-cardboard interaction, around 50% for jet-metal interaction, and approximately 200% for jet-DI water interaction, respectively. This change in optical emissions spectrometry (OES) intensity for DI water and metal, but not cardboard, implies a change in the plasma plume characteristics when certain subjects are placed nearer to the nozzle opening, likely related to the electrical conductivity of the subjects. While DI water is non-conductive itself, it becomes conductive after plasma treatment [29]. Thus, DI water and metal as conductive subjects result in higher plasma intensity than the cardboard as a non-conductive subject. Plasma interacting with DI water will generate ROS/RNS, such as hydrogen peroxide, nitrite, and other species, in the water to form a plasma solution. In our previous papers, we indicated that the concentration of ROS/RNS depends on the discharge voltage/current, types of feeding gas, feeding gas flow rate, water volume, and other parameters [3,19,25,29]. For the same parameters, the concentration of ROS/RNS in water exhibits time-dependent behavior. In addition, the pH of the plasma solution will increase with the increasing time of plasma-water interaction [14]. Interaction between the tiny plasma jet and the metal surface may bias the surface positively and, in addition to standard fluid motion, cause the discharge to spread out horizontally along the metal's surface further than with other materials. At the interaction between the plasma jet and DI water, the pressure from the plasma flow creates a small pocket in the water's surface, potentially modifying the local airflow to reduce the dispersion of reactive species along the surface; along with the tendency of jet/water interactions to increase reactive species generation, this discharge counterflow may help to increase species density near the emission tip. This may explain why the OES intensity of the jet interacting with DI water is measured as higher than when interacting with metal. Gerling et al. indicated that the distance between the electrode and the ground is significant in generating different plasma behaviors, with a "pre-bullet" being generated at a 15 mm gap but no "pre-bullet" at 10 mm [30]. Since we're using a single electrode in this situation, the subject might be considered a floating electrode and, for the 15 mm gap, what we might be seeing is the incomplete discharge for the pre-bullet, resulting in the decreased OES intensities. Nastuta et al. tested atmospheric pressure at the plasma jet-living tissue interface at 5 mm and 15 mm and found similar drops in intensity for subjects [31].

Figure 5a shows images of the tiny plasma jet interacting with pork belly and muscle at 10 mm and 15 mm distance. From 15 mm to 10 mm, the plasma intensity at both subjects increase, and the OES results in Figure 5b also confirm this. Comparing OES results for belly and muscle at 15 mm distance with the tiny plasma jet alone, there is minimal change in ROS/RNS intensity, while ROS/RNS intensity largely increases when the distance is reduced from 15 mm to 10 mm. In addition, ROS/RNS intensity of jet-muscle interaction increases more than jet-belly interaction at the 10 mm distance. Figure 5c shows high-resolution images of pork belly and muscle before and after 2 min tiny plasma treatment at 10 mm and 15 mm distance. There is no damage or burning on belly and muscle surfaces after 2 min treatment at 10 mm and 15 mm. We tried to reduce the distance and increase other inputs to see the effect on biological tissues. The results show no damage to tissues. The reason might be that lower input power is too low to damage tissues. Figure 6 exhibits thermal images of the tiny plasma jet interacting with five subjects after 2 min at 10 mm and 15 mm. For cardboard and DI water, the center temperatures of the interacting area decrease from 15 mm to 10 mm; however, the variations are small (0.6 °C and 0.5 °C, respectively) and may be considered noise. Due to jet flow, indentations form on the water surface at both distances, which may partly explain why ROS/RNS intensities are higher than with other subjects. For metal, belly, and muscle, the center temperature after 2 min interaction increases when the distance is changed from 15 mm to 10 mm (increases of 2.2 °C, 3.0 °C, and 6.6 °C, respectively), but it still remains at a comfortable temperature for human beings. The belly and muscle samples, in particular, begin at significantly cooler temperatures, and the interaction with the gas alone from the jet would be expected to increase the temperature of the target area to room temperature, as can be seen in the larger outer thermal "rings" in Figure 6, which are near-universal in size and temperature between non-liquid subjects; the more relevant interaction with the plasma discharge for the belly and muscle samples can be identified by the central thermal hotspots. Thus, the tiny plasma jet should be an ideal and safe tool for in vivo biomedical applications.

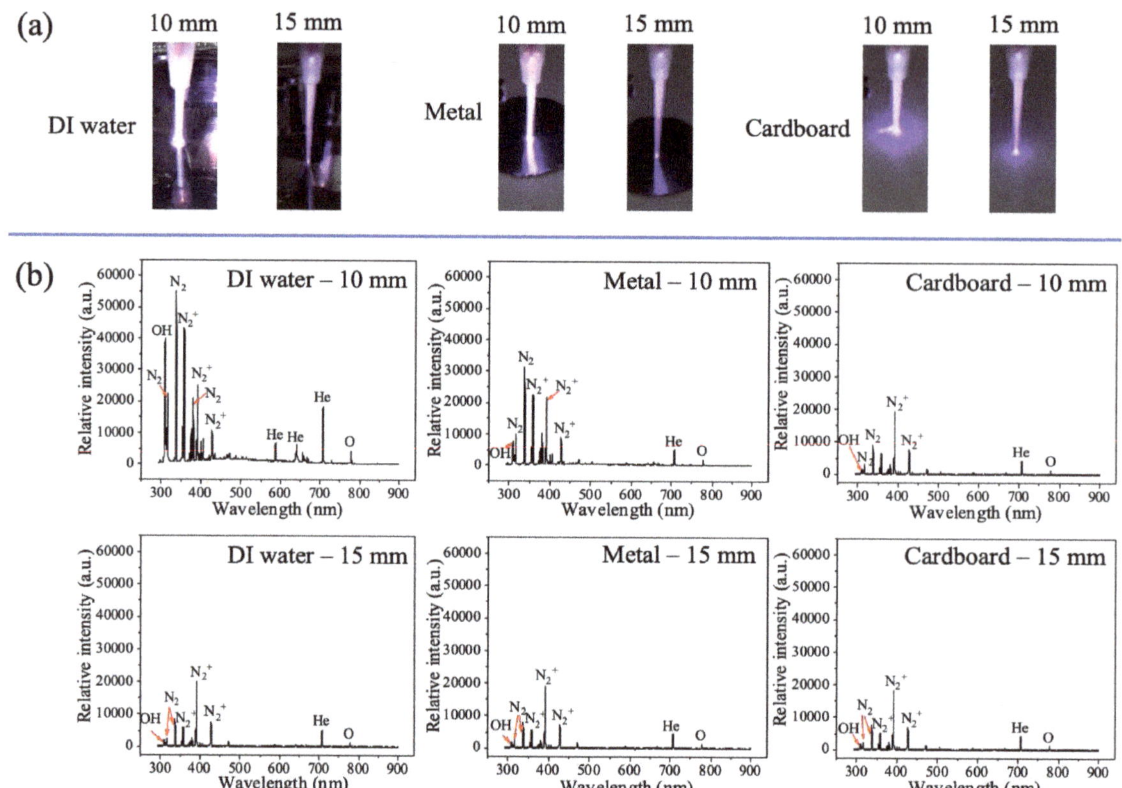

Figure 4. (**a**) Images of the tiny plasma jet interacting with deionized (DI) water, metal, and cardboard at 10 mm and 15 mm distance. (**b**) The effect of interacting processes between jets and subjects (DI water, metal, and cardboard) on optical emission spectra of tiny plasma jets.

Cold plasma has received considerable attention for its potential biomedical applications, including blood coagulation, wound healing, sterilization/inactivation, skin regeneration, oral health, and cancer therapy [4,32,33]. The efficiency of plasma for biomedical applications mainly relies on the ROS/RNS generated in it [34,35]. As shown in Figures 2d, 4b and 5b, the tiny plasma jet is an ideal ROS/RNS source, and ROS/RNS intensity increases when the plasma jet is used nearer to tissues and conductive materials. This single-electrode tiny plasma jet without the major drawbacks of conventional plasma jet (high voltage, high flow rate, reactive species delivery, plasma jet volume, and discharge formation in the organ) will have more potential biomedical applications, especially for in vivo clinical applications. For example, plasma disinfects bacteria in the oral cavity, such as *Enterococcus faecalis* and *Candida albicans* [36,37]. This reliable and user-friendly tiny plasma jet could be directed manually to target a root canal for disinfection. A tiny plasma jet is also an ideal tool when applied inside a living body, especially for cancer therapy. We take a glioblastoma, for example, which is a highly malignant aggressive neoplasm with rapid growth and resistance to all current therapies [38,39]. Our previous results have shown that plasma is effective at preventing glioblastoma growth in the mouse brain [40]. The tiny plasma jet, with a low flow rate and limited discharge voltage, can be generated in closer proximity to the organ, safely delivering ROS/RNS across the blood-brain barrier to the tumor. Overall, the tiny plasma jet is useful and should be considered in clinical medical applications. A further understanding of the precise underlying mechanisms will provide the 'best' combination when employed as a treatment strategy.

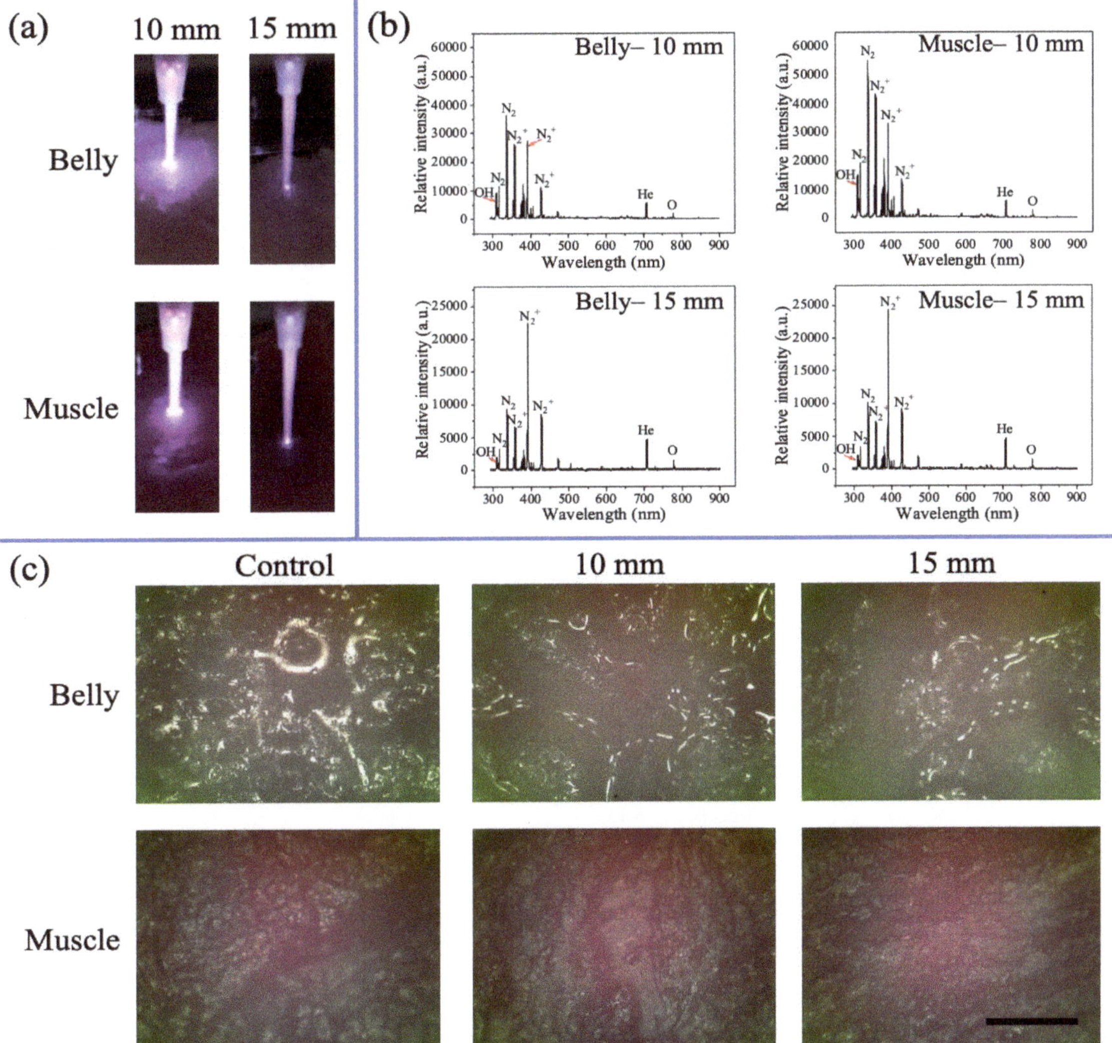

Figure 5. (**a**) Images of tiny plasma jet interacting with pork belly and muscle at 10 mm and 15 mm distance. (**b**) The effect of interacting processes between jets and pork (belly and muscle) on optical emission spectra of tiny plasma jets. (**c**) The surface images of pork (belly and muscle) after jet-pork interactions for control (0 min), 10 mm (2 min), and 15 mm (2 min). Tissue surface measurements were made using a BW500 Digital Microscope to examine the surface of each sample. Scale bar: 1 mm.

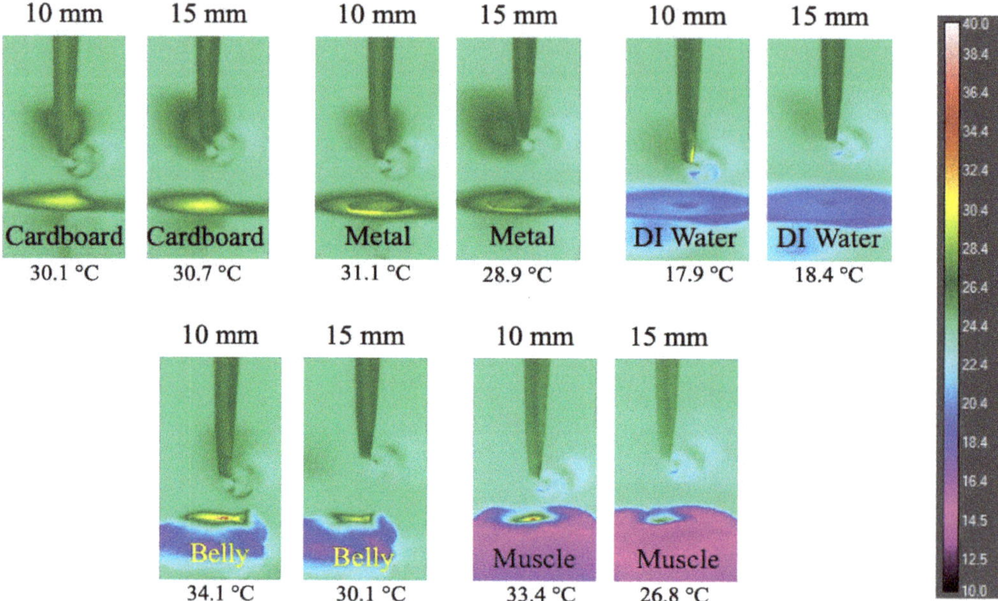

Figure 6. Thermal measurement of the tiny plasma jet interacting with cardboard, metal, DI water, belly, and muscle for 2 min at 10 mm and 15 mm distances. The temperature marked below each image is the center temperature of a tiny plasma jet interacting with subjects.

4. Conclusions

In summary, we developed a single-electrode tiny plasma device and investigated its physics and interactions with five subjects (DI water, metal, cardboard, belly, and muscle) at 10 mm and 15 mm. For non-conductive subjects, ROS/RNS intensity shows very little change when the distance is decreased from 15 mm to 10 mm, while ROS/RNS intensity increases for conductive subjects with distance decreasing from 15 mm to 10 mm, especially for the muscle. For five subjects, the center temperature of jet-subjects interaction still remains in the comfortable temperature range for human beings after 2 min interaction for both 10 mm and 15 mm distances. High-resolution images of belly and pork after 2 min tiny plasma treatment show no damage or burning on tissues' surfaces. The results of this study are preliminary research for employing the tiny plasma jet in clinical biomedical applications in the future.

Author Contributions: Z.C. and R.O.: Conceptualization, methodology, experiments, formal analysis, resources, project administration, writing; R.E.W.: Conceptualization, methodology, project administration, writing. All authors have read and agreed to the published version of the manuscript.

Funding: This work was supported by the UCLA Plasma and Space Propulsion Laboratory.

Institutional Review Board Statement: Not Applicable.

Informed Consent Statement: Not Applicable.

Data Availability Statement: The data that support the findings of this study are available from the corresponding author upon reasonable request.

Conflicts of Interest: The authors declare no conflict of interest.

References

1. Morozov, A.I. *Introduction to Plasma Dynamics*; CRC Press: Boca Raton, FL, USA, 2012.
2. Lu, X.; Naidis, G.; Laroussi, M.; Reuter, S.; Graves, D.; Ostrikov, K. Reactive species in non-equilibrium atmospheric-pressure plasmas: Generation, transport, and biological effects. *Phys. Rep.* **2016**, *630*, 1. [CrossRef]
3. Chen, Z. Development of New Cold Atmospheric Plasma Devices and Approaches for Cancer Treatment. Ph.D. Thesis, The George Washington University, Washington, DC, USA, 2018.
4. Fridman, G.; Gutsol, A.; Shekhter, A.B.; Vasilets, V.N.; Fridman, A. Applied Plasma Medicine. *Plasma Process. Polym.* **2008**, *5*, 503. [CrossRef]
5. Li, W.; Yu, H.; Ding, D.; Chen, Z.; Wang, Y.; Wang, S.; Li, X.; Keidar, M.; Zhang, W. Cold atmospheric plasma and iron oxide-based magnetic nanoparticles for synergetic lung cancer therapy. *Free Radic. Biol. Med.* **2019**, *130*, 71. [CrossRef] [PubMed]
6. Kong, M.G.; Kroesen, G.; Morfill, G.; Nosenko, T.; Shimizu, T.; van Dijk, J.; Zimmermann, J. Plasma medicine: An introductory review. *New J. Phys.* **2009**, *11*, 115012. [CrossRef]
7. Laroussi, M. Sterilization of contaminated matter with an atmospheric pressure plasma. *IEEE Trans. Plasma Sci.* **1996**, *24*, 1188. [CrossRef]
8. Isbary, G.; Morfill, G.; Schmidt, H.; Georgi, M.; Ramrath, K.; Heinlin, J.; Karrer, S.; Landthaler, M.; Shimizu, T.; Steffes, B. A first prospective randomized controlled trial to decrease bacterial load using cold atmospheric argon plasma on chronic wounds in patients. *Br. J. Dermatol.* **2010**, *163*, 78.
9. Pan, J.; Sun, P.; Tian, Y.; Zhou, H.; Wu, H.; Bai, N.; Liu, F.; Zhu, W.; Zhang, J.; Becker, K.H. A novel method of tooth whitening using cold plasma microjet driven by direct current in atmospheric-pressure air. *IEEE Trans. Plasma Sci.* **2010**, *38*, 3143. [CrossRef]
10. Chen, G.; Chen, Z.; Wen, D.; Wang, Z.; Li, H.; Zeng, Y.; Dotti, G.; Wirz, R.E.; Gu, Z. Transdermal cold atmospheric plasma-mediated immune checkpoint blockade therapy. *Proc. Natl. Acad. Sci. USA* **2020**, *117*, 3687. [CrossRef]
11. Joshi, S.G.; Cooper, M.; Yost, A.; Paff, M.; Ercan, U.K.; Fridman, G.; Friedman, G.; Fridman, A.; Brooks, A.D. Nonthermal dielectric-barrier discharge plasma-induced inactivation involves oxidative DNA damage and membrane lipid peroxidation in Escherichia coli. *Antimicrob. Agents Chemother.* **2011**, *55*, 1053. [CrossRef]
12. Schmidt, A.; Bekeschus, S.; Wende, K.; Vollmar, B.; von Woedtke, T. A cold plasma jet accelerates wound healing in a murine model of full-thickness skin wounds. *Exp. Dermatol.* **2017**, *26*, 156.
13. Keping, Y.; Qikang, J.; Zheng, C.; Guanlei, D.; Shengyong, Y.; Zhen, L. Pulsed cold plasma-induced blood coagulation and its pilot application in stanching bleeding during rat hepatectomy. *Plasma Sci. Technol.* **2018**, *20*, 044005.
14. Chen, Z.; Lin, L.; Cheng, X.; Gjika, E.; Keidar, M. Effects of cold atmospheric plasma generated in deionized water in cell cancer therapy. *Plasma Process. Polym.* **2016**, *13*, 1151. [CrossRef]
15. Khlyustova, A.; Labay, C.; Machala, Z.; Ginebra, M.-P.; Canal, C. Important parameters in plasma jets for the production of RONS in liquids for plasma medicine: A brief review. *Front. Chem. Sci. Eng.* **2019**, *13*, 238. [CrossRef]
16. Chen, Z.; Garcia, G., Jr.; Arumugaswami, V.; Wirz, R.E. Cold atmospheric plasma for SARS-CoV-2 inactivation. *Phys. Fluids* **2020**, *32*, 111702. [CrossRef] [PubMed]
17. Chen, Z.; Xu, R.-G.; Chen, P.; Wang, Q. Potential Agricultural and Biomedical Applications of Cold Atmospheric Plasma-Activated Liquids with Self-Organized Patterns Formed at the Interface. *IEEE Trans. Plasma Sci.* **2020**, *48*, 3455. [CrossRef]
18. Graves, D.B. The emerging role of reactive oxygen and nitrogen species in redox biology and some implications for plasma applications to medicine and biology. *J. Phys. D Appl. Phys.* **2012**, *45*, 263001. [CrossRef]
19. Chen, Z.; Cheng, X.; Lin, L.; Keidar, M. Cold atmospheric plasma discharged in water and its potential use in cancer therapy. *J. Phys. D Appl. Phys.* **2016**, *50*, 015208. [CrossRef]
20. Xu, R.-G.; Chen, Z.; Keidar, M.; Leng, Y. The impact of radicals in cold atmospheric plasma on the structural modification of gap junction: A reactive molecular dynamics study. *Int. J. Smart Nano Mater.* **2019**, *10*, 144. [CrossRef]
21. Mirpour, S.; Piroozmand, S.; Soleimani, N.; Faharani, N.J.; Ghomi, H.; Eskandari, H.F.; Sharifi, A.M.; Mirpour, S.; Eftekhari, M.; Nikkhah, M. Utilizing the micron sized non-thermal atmospheric pressure plasma inside the animal body for the tumor treatment application. *Sci. Rep.* **2016**, *6*, 29048. [CrossRef]
22. Shashurin, A.; Keidar, M. Experimental approaches for studying non-equilibrium atmospheric plasma jets. *Phys. Plasmas* **2015**, *22*, 122002. [CrossRef]
23. Pearse, R.W.B.; Gaydon, A.G. *Identification of Molecular Spectra*; Chapman and Hall: Boca Raton, FL, USA, 1976.
24. Cheng, X.; Sherman, J.; Murphy, W.; Ratovitski, E.; Canady, J.; Keidar, M. The Effect of Tuning Cold Plasma Composition on Glioblastoma Cell Viability. *PLoS ONE* **2014**, *9*, e98652. [CrossRef] [PubMed]
25. Chen, Z.; Lin, L.; Cheng, X.; Gjika, E.; Keidar, M. Treatment of gastric cancer cells with nonthermal atmospheric plasma generated in water. *Biointerphases* **2016**, *11*, 031010. [CrossRef] [PubMed]
26. Xiong, Q.; Lu, X.; Ostrikov, K.; Xiong, Z.; Xian, Y.; Zhou, F.; Zou, C.; Hu, J.; Gong, W.; Jiang, Z. Length control of He atmospheric plasma jet plumes: Effects of discharge parameters and ambient air. *Phys. Plasmas* **2009**, *16*, 043505. [CrossRef]
27. Li, Q.; Li, J.-T.; Zhu, W.-C.; Zhu, X.-M.; Pu, Y.-K. Effects of gas flow rate on the length of atmospheric pressure nonequilibrium plasma jets. *Appl. Phys. Lett.* **2009**, *95*, 141502. [CrossRef]
28. Lu, X.; Keidar, M.; Laroussi, M.; Choi, E.; Szili, E.; Ostrikov, K. Transcutaneous plasma stress: From soft-matter models to living tissues. *Mater. Sci. Eng. R Rep.* **2019**, *138*, 36. [CrossRef]

29. Chen, Z.; Zhang, S.; Levchenko, I.; Beilis, I.I.; Keidar, M. In vitro demonstration of cancer inhibiting properties from stratified self-organized plasma-liquid interface. *Sci. Rep.* **2017**, *7*, 12163. [CrossRef] [PubMed]
30. Gerling, T.; Nastuta, A.; Bussiahn, R.; Kindel, E.; Weltmann, K. Back and forth directed plasma bullets in a helium atmospheric pressure needle-to-plane discharge with oxygen admixtures. *Plasma Sour. Sci. Technol.* **2012**, *21*, 034012. [CrossRef]
31. Nastuta, V.; Pohoata, V.; Topala, I. Atmospheric pressure plasma jet—Living tissue interface: Electrical, optical, and spectral characterization. *J. Appl. Phys.* **2013**, *113*, 183302. [CrossRef]
32. Babaeva, N.Y.; Naidis, G.V. Modeling of plasmas for biomedicine. *Trends Biotechnol.* **2018**, *36*, 603. [CrossRef]
33. Laroussi, M.; Lu, X.; Keidar, M. Perspective: The physics, diagnostics, and applications of atmospheric pressure low temperature plasma sources used in plasma medicine. *J. Appl. Phys.* **2017**, *122*, 020901.
34. Labay, C.; Roldán, M.; Tampieri, F.; Stancampiano, A.; Bocanegra, P.E.; Ginebra, M.-P.; Canal, C. Enhanced Generation of Reactive Species by Cold Plasma in Gelatin Solutions for Selective Cancer Cell Death. *ACS Appl. Mater. Interfaces* **2020**, *12*, 47256–47269. [CrossRef] [PubMed]
35. Szili, E.J.; Hong, S.-H.; Oh, J.-S.; Gaur, N.; Short, R.D. Tracking the penetration of plasma reactive species in tissue models. *Trends Biotechnol.* **2018**, *36*, 594. [CrossRef] [PubMed]
36. Lu, X.; Cao, Y.; Yang, P.; Xiong, Q.; Xiong, Z.; Xian, Y.; Pan, Y. An RC Plasma Device for Sterilization of Root Canal of Teeth. *IEEE Trans. Plasma Sci.* **2009**, *37*, 668.
37. Kerlikowski, A.; Matthes, R.; Pink, C.; Steffen, H.; Schlüter, R.; Holtfreter, B.; Weltmann, K.D.; von Woedtke, T.; Kocher, T.; Jablonowski, L. Effects of cold atmospheric pressure plasma and disinfecting agents on Candida albicans in root canals of extracted human teeth. *J. Biophotonics* **2020**, *13*, e202000221. [CrossRef]
38. Lim, M.; Xia, Y.; Bettegowda, C.; Weller, M. Current state of immunotherapy for glioblastoma. *Nat. Rev. Clin. Oncol.* **2018**, *15*, 422. [CrossRef]
39. Wick, W.; Gorlia, T.; Bendszus, M.; Taphoorn, M.; Sahm, F.; Harting, I.; Brandes, A.A.; Taal, W.; Domont, J.; Idbaih, A. Lomustine and bevacizumab in progressive glioblastoma. *N. Engl. J. Med.* **2017**, *377*, 1954. [CrossRef]
40. Chen, Z.; Simonyan, H.; Cheng, X.; Gjika, E.; Lin, L.; Canady, J.; Sherman, J.H.; Young, C.; Keidar, M. A novel micro cold atmospheric plasma device for glioblastoma both in vitro and in vivo. *Cancers* **2017**, *9*, 61. [CrossRef]

Article

The Combination of Simultaneous Plasma Treatment with Mg Nanoparticles Deposition Technique for Better Mung Bean Seeds Germination

Sarunas Varnagiris [1,*], Simona Vilimaite [2], Ieva Mikelionyte [2], Marius Urbonavicius [1], Simona Tuckute [1] and Darius Milcius [1]

1 Center for Hydrogen Energy Technologies, Lithuanian Energy Institute, 3 Breslaujos, 44403 Kaunas, Lithuania; marius.urbonavicius@lei.lt (M.U.); simona.tuckute@lei.lt (S.T.); darius.milcius@lei.lt (D.M.)

2 Gymnasium of Lithuanian University of Health Sciences, LT-47182 Kaunas, Lithuania; simonavilimaite01@gmail.com (S.V.); ievamikel@gmail.com (I.M.)

* Correspondence: sarunas.varnagiris@lei.lt; Tel.: +370-37-401-824

Received: 29 October 2020; Accepted: 27 November 2020; Published: 29 November 2020

Abstract: A novel method based on the combination of simultaneous cold plasma treatment with Mg nanoparticles deposition, applied to Mung bean seeds by improving their quality, is presented. The SRIM simulation reveals that only the very top layer of the seeds surface can be altered by the plasma. The experimental analysis indicates surface composition changes with a polar groups formation. These groups initiate the shift of surface characteristics from hydrophobic to hydrophilic. The chemical bond analysis shows the formation of MgO and $Mg(OH)_2$ compounds, which acts as a positive factor for seeds germination and growth. The germination experiments showed a 70% outcome with an average of 73.9 mm sprouts length after 30 min of plasma treatment compared to the initial seeds (40% outcome and 71.3 mm sprouts length).

Keywords: germination; Mg deposition; non-thermal plasma; plasma treatment; surface modification

1. Introduction

Mung beans are an important crop widely cultivated across Asia and some parts of America [1]. According to the IMARC group (market forecasters), the global market of Mung beans reached an amount of 3.4 million tons in 2019 where India is a leader with 60% of the total production. The Mung beans have gained their popularity as edible sprouts due to the numerous health benefits and their high nutritional value, which include proteins, fibers, vitamins, minerals, antioxidants, and phytoestrogens [2,3]. Since the demand of sprouts is still increasing, the crop yield and rate of seeds germination should increase, respectively.

However, some factors and an insufficient germination of seeds could lead to a reduction in the production yield of Mung bean sprouts. The main causes of the low germination of plant seeds are usually connected to genetic factors, the quality of the seeds surface and soil contamination with microorganisms, as well as the lack of moisture or macronutrients [4,5]. Moreover, adverse environmental conditions can negatively affect the germination.

The conventional methods used to enhance the germination rate are fertilization and increased irrigation but it still faces economic and environmental issues [4,6]. The agrochemicals used for yield improvement might be hazardous for the human health [7]. Moreover, current consumers are interested in less chemically treated food products. Alternatively, genetic engineering, seeds sterilization with sodium chlorite, treatment with acids, antibiotics, hot water, or additional macro elements can be employed for better germination [1,8,9]. However, these processes are considered complex, expensive, not always efficient, and sometimes even harmful. Therefore, eco-friendly and non-hazardous alternative methods are desired.

It is noted that seed germination begins with a water uptake, which induces biochemical processes [5,10]. The absorption of the water rate is influenced by environmental and plant seed surface properties (morphological and chemical state). Although increased surface wettability and water uptake are necessary, they are not sufficient conditions for better germination.

Recently, the treatment of plant seeds surface using non-thermal plasma is one of the emerging technologies in the agriculture, which can breakdown seed dormancy, control water absorption, kill bacteria, or positively stimulate germination and growth [11]. Consequently, it gets more and more attention in China, which uses plasma treating seed technologies commercialized by Russia [12]. However, the existing devices still need many improvements.

The non-thermal plasma consisting of ions, electrons, excited atoms, radicals, molecules, and UV radiation can modify the surface from several up to tenths of nanometers [13]. During the non-thermal plasma treatment, plant seeds undergo only a very low stress via plasma activation and are kept intact. Due to such characteristics, the plasma can induce a mild surface etching (via ions bombardment) or even enrich the seeds surface with oxygen containing functional groups [5]. These groups can significantly increase the surface wettability, which has a very positive effect on seeds metabolism and germination, as well as surface permeability for nutrients [3,7]. Another positive effect of plasma modification is the sterilization of the seeds surface and elimination of unwanted microbes [1,14,15]. Moreover, the non-thermal plasma is suitable for heat sensitive surfaces, as well as being a fast and environmental-friendly (without any hazardous chemicals) method, which provides a uniform and non-destructive surface treatment of plant seeds [7,16]. It is demonstrated that various treatment parameters can accelerate germination, increase sprout growth, plant height, weight or root growth, decontaminate the seed surface, or improve the survivability [17]. However, the process mechanism can still be investigated in more detail.

Among different common gases (such as Ar, He, N, O_2), very promising results are shown after the exposure of seeds to the atmospheric air plasma where enhanced germination and seedling vigor were observed [17–20]. In the air plasma, reactive nitrogen and oxygen species are generated (RONS and ROS, respectively). Furthermore, the air plasma consists of predominant excited oxygen (O_2*) and nitrogen (N_2*) molecules, atomic nitrogen (N) and oxygen (O), superoxide anions (O_2^-), and some radicals such as, H_2O^+, OH^-, and OH* [21,22]. The reactive oxygen and nitrogen species have a huge impact on seeds activation in the plasma [8,23].

Nevertheless, the impact of micro nutrients on seeds germination should not be forgotten. Magnesium plays an important role in plants as a regulator of biochemical functions and can have a positive impact on seeds germination, as well as plant growth [24]. It activates enzymes and is an essential element in the photosynthesis process. Moreover, it is a very important nutrient for human health [24,25]. S. Shinde et al. synthesized magnesium hydroxide ($Mg(OH)_2$) nanoparticles and investigated their influence on the germination of *Zea mays* [26]. The authors found a significant increase in seeds germination and growth (i.e., enhanced shoot height and root length). Another research group demonstrated that magnesium oxide nanoparticles also improved the green gram (*V. radiata*) seeds germination performance [27]. Generally, the plasma treatment of plants and seeds as well as the usage of additional nutrients are separate approaches and studies for an increment of germination efficiency. Magnesium is one of the most important elements for the healthy growth of plants and one of the most limiting macronutrients in agriculture [26,28].

This study focused on the theoretical and experimental analysis of Mung bean seeds surface modifications, by an implanted gaseous species from the plasma and nanoparticles, which arrive from magnesium cathodes used during the cold plasma treatment. The oxygen and nitrogen implantation could promote beans germination and initial growth. Moreover, magnesium nano clusters could act as nano fertilizers, which improve the plant nutrition level during growth. Basically, magnesium salts are used as fertilizers in order to increase their concentration in soil. Here, we show an original approach attaching magnesium nanoparticles directly to the seeds. Specifically, we investigate the influence of the treatment on the chemical state characteristics of the seeds surface, the change in the elemental composition, as well as germination by applying various surface analysis techniques. In this study, we demonstrate that a combination of simultaneous non-thermal plasma treatment with Mg nanoparticles deposition on the surface have a significant potential for improving the outcome of seeds germination.

2. Experimental Details

2.1. Plasma Treatment

During the plasma treatment, the Mung bean seeds were placed in a vacuum chamber on metal mesh between two Mg electrodes (produced by KJLC, Clairton, PA, USA 99.9% purity). The principle scheme of the plasma treatment as well as the image of the real plasma treatment process is shown in Figure 1. The two electrodes and metal mesh between them was used in order to ensure a homogeneous treatment process for all the seeds surface. The direct current power source (20 W) was used for plasma generation, while air was used as a working gas. The pressure of 10 Pa was kept constant during the plasma treatment process (gas flow was maintained around 30 mL/min). The distance between each Mg electrode and metal mesh was 4 cm. Mung bean seeds were treated in plasma for 10, 30, 60, and 90 min.

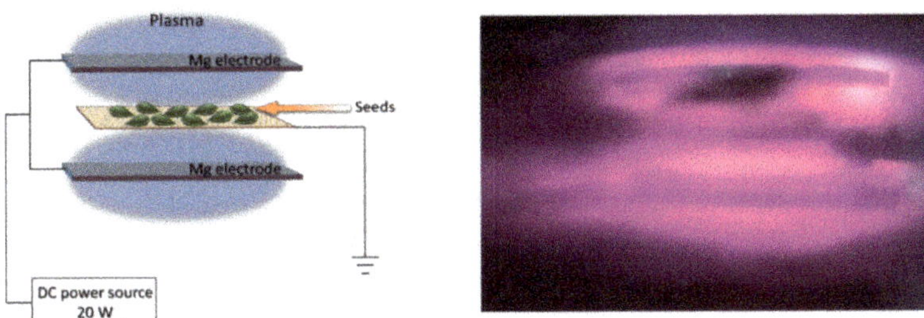

Figure 1. Experimental setup of seeds plasma treatment.

2.2. Plasma Treatment Process Simulations

The air-plasma ion penetration into the Mung bean seeds was evaluated by using SRIM—a free available Monte Carlo simulation code with a full name: Stopping and Range of Ions in Materials. SRIM is the most widespread simulation code mainly used for calculations of sputtering and its application areas. The SRIM software is based on the TRIM code (TRansport of Ions in Matter), which uses Biersack's magic formula and the ZBL (Ziegler-Biersack-Littmark) universal interaction potential for its calculations [29,30].

2.3. Characterization

The surface wettability was evaluated by water contact angle (WCA) measurements, which were performed immediately (in less than 5 min) after the plasma treatment process. The laboratory made equipment with a camera and PC computer was used for high resolution pictures of the water drop on the seeds surface. Deionized water was syringed on the surface of the seeds and the pictures were taken during the next 3 s. Three seeds were used for each measurement and averaged values were calculated. The surface elemental composition and chemical bond analysis of the plasma treated Mung bean seeds were analyzed by an X-ray photoelectron spectroscope (XPS, PHI 5000 Versaprobe, Chanhassen, MN, USA). During the XPS analysis, the monochromated 1486.6 eV Al radiation, 25 W beam power, 100 μm beam size, and 45° were used. The Multipak software and NIST Standard Reference Database were used for the XPS spectra processing and analysis. The elemental mapping of Mung bean seeds was done using a scanning electron microscope (Hitachi S-3400N, Tokyo, Japan) equipped with energy dispersive X-ray spectroscopy (EDS, Bruker Quad 5040, Berlin, Germany).

2.4. Germination

The germination rate is presented as a percentage by indicating the proportion of germinated and total number of Mung bean seeds, which were seeding for 7 days at room temperature (the temperature was kept between 21–23 °C during the whole germination process) with 12 h light and 12 h dark. Twenty Mung bean seeds of each group (initial, 10, 30, 60, and 90 min of the plasma treatment) were used for the germination experiments. The humidity was set between 60–70%. The calculation of the germination rate is presented by the following equation:

$$Germination\ rate\ (\%) = \left(\frac{Number\ of\ germinated\ seeds}{Total\ number\ of\ seeds}\right) \times 100\ \% \quad (1)$$

3. Results and Discussion

3.1. Simulation Results

During the SRIM simulation, some assumptions were taken into account, therefore, the real penetration depth might vary by several nanometers. First, it was assumed that the surface of the Mung bean seeds consists of carbon only. The elemental analysis experiments, which are presented below, revealed that the surface of the Mung bean seeds consist of carbon (more than 80%) and some other elements: O, Si, and Ca. This result was observed by analyzing a very thin surface layer, which was up to 10 nm. Moreover, it is known that the carbon concentration could even increase to a very top layer due to the adventitious carbon [31]. Therefore, the simulation was performed by approaching the plasma ions that penetrate through the carbon layer of the analyzed seeds and using the density of carbon during the simulation. Second, nitrogen is a main component of air, which was used for the plasma treatment. Therefore, the simulation was simplified by taking into account nitrogen as an ion source. Finally, the calculated air plasma as well as the Mg ion energy, which are required for the simulation, were about 0.1–0.2 keV.

Figure 2 shows the simulation results of the air-plasma and Mg ions penetration in to the Mung bean seeds. The results show that the air-plasma ions penetrated into approximately 3 nm depth of the Mung bean seeds, while the highest concentration of plasma ions penetration was fixed at a 2 nm depth. This means that only a very top layer of the seeds surface is altered during the plasma treatment and such a method does not make any changes on the seeds bulk. However, a majority of the researchers claimed that the plasma ion penetration depth into the various seeds is up to 10 nm [10,32,33]. This disparity between the calculated results and the ion penetration depth provided by other authors could be related to several approaches. First, a majority of the other authors used the term "up to 10 nm" without any ion penetration depth calculations, while it is widely known that the plasma can influence only the "very top layer" of the treated material. Up to our knowledge, there are no articles that indicate a particular penetration depth of cold plasma ions into the seed. Moreover, the existing ion penetration depth models into other materials used different plasma parameters including ion flux, temperature, plasma composition, etc. Therefore, it is hard to evaluate the real difference between our suggested model and the penetration depth results presented by the other authors.

Moreover, our suggested model revealed that the deposited Mg nanoparticles could penetrate into 1–2 nm of the Mung bean surface (Figure 2b). The result that air-plasma ions can penetrate deeper than Mg is reasonable, while both air-plasma and Mg ions have a similar energy during the plasma treatment and air-plasma ions have a lower diameter, determining an easier penetration process.

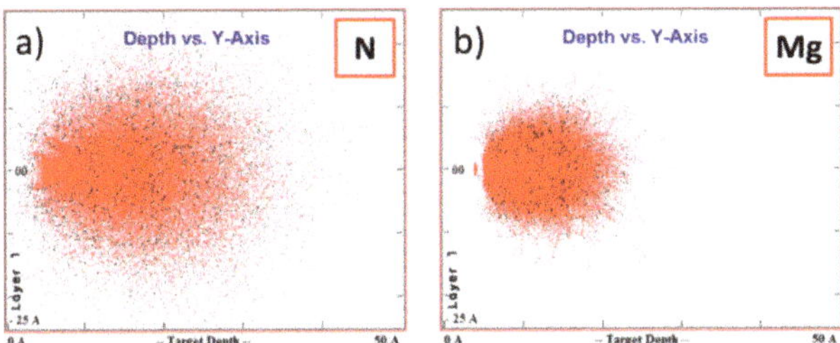

Figure 2. Simulation results of (**a**) air-plasma and (**b**) Mg ions penetration in to the Mung bean seeds: Visualization of ions distribution.

The air plasma consists of a variety of different particles and radicals (electrons, neutral, particles, excited atoms, molecules), as well as UV irradiation. Therefore, it is well known that the gas plasma can result in various effects upon the treated surface including cleaning or alteration of the surface. Many processes can occur during the interaction of air plasma with the seeds surface: The removal of surface contamination, the formation of an altered layer by the implantation of magnesium ions and active ions from air plasma near the surface, the adsorption of magnesium atoms on the surface, the formation of polar molecule groups on the surface which promote surface wettability, the changes in the surface topography on the nanoscale, etc. [8,34]. Under certain conditions, magnesium particles from the used electrodes are sputtered by ions existing in the air plasma. The sputtered magnesium with enough energy is deposited on the seeds surface and oxidized by an active oxygen species in the air plasma (Figure 3).

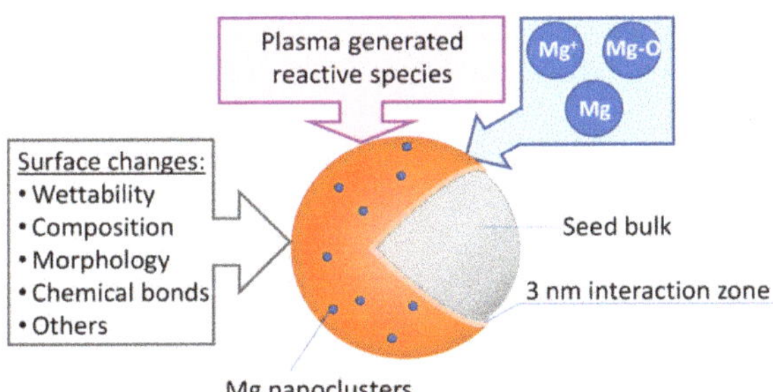

Figure 3. The scheme of the plasma treatment process with Mg nanoclusters deposition on the seed surface.

3.2. Experimental Results

An elemental analysis of the very top layer of initial and plasma-treated Mung bean seeds was performed by XPS measurement (Figure 4). The surface of the initial seeds consists of carbon (80.5 at%), oxygen (15.3 at%), nitrogen (2.0 at%) and a small amount of trace elements (2.2 at%). The clear changes of the Mung bean seeds surface elemental composition were observed after the plasma treatment. The amount of carbon decreased approximately two times even after 10 min of the plasma treatment (43.7 at%). A longer treatment time invoked the reduction of carbon which varied from 24 to 34.7 at%. On the contrary, the increment of oxygen was observed after the first 10 min of the plasma treatment up to 42.3 at%. Such an amount of oxygen remained relatively stable despite a longer treatment time. Similar tendencies were observed by estimating the amount of nitrogen, in which the concentration slightly increased after the plasma treatment and varied in a narrow range (from 2.4 to 3.6 at%).

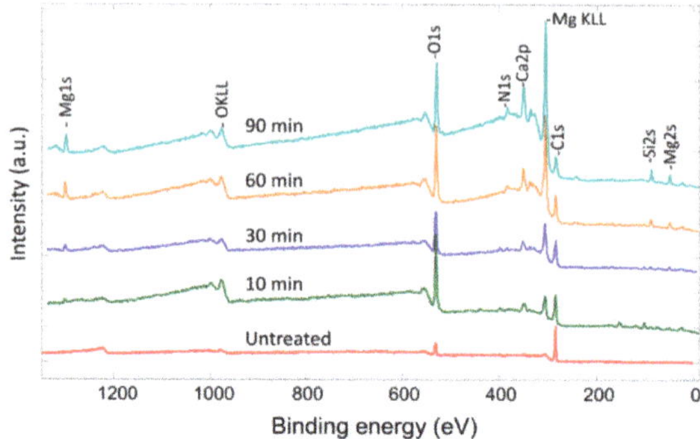

Figure 4. XPS survey spectra of initial- and plasma-treated Mung bean seeds.

These results are determined by the native characteristics of the plasma, while two main processes (known as plasma cleaning and free radical sites formation) complement each other during the plasma treatment. The plasma cleaning initiates the removal of organic contaminants, as well as an adventitious carbon from the surface of the sample. Such a process can be related to the carbon amount reduction, as well as the oxygen promotion during the plasma treatment. Meanwhile, the formation of free radical sites usually improves the free surface energy and induces a re-orientation with new chemical bonds formation (see Figure 5) [35–38].

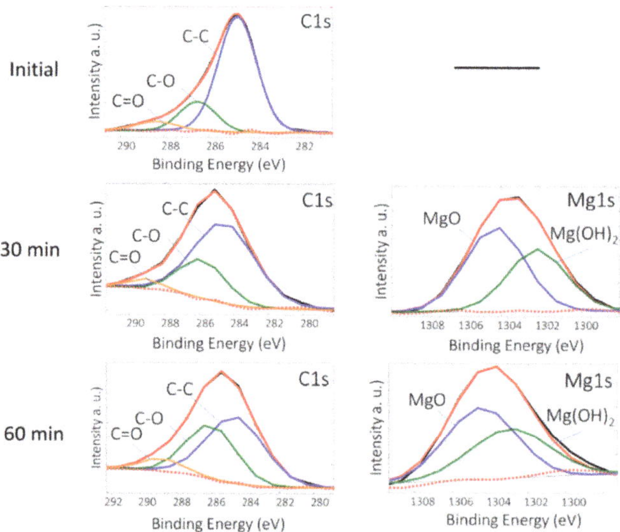

Figure 5. XPS fitting results of C1s and Mg1s peaks.

One of the main aims of this work was to form Mg nanoclusters on the surface of Mung bean seeds during the plasma treatment. The analysis showed that 2.9 at% of Mg was observed even after 10 min of the plasma treatment. Moreover, the concentration of Mg increased with a longer plasma treatment time and reached 19.0 at% after 90 min of the plasma treatment. This result confirms that Mg nanoparticles were deposited on the surface of Mung bean seeds during the plasma treatment. The small amount of trace elements (Si, Ca), which is a component of the seed sheath, was observed by analyzing all types of seeds. The small increase of Si and Ca elements after the plasma treatment compared to the initial seeds concentration might be related to the plasma cleaning process, where organic contaminants are removed and a more certain elemental composition of seed sheath could be observed. The detailed results of the Mung bean surface elemental composition before and after the plasma treatment are presented in Table 1.

Table 1. Concentrations of elemental composition (up to 10 nm) before and after the plasma treatment.

Samples	Concentration, at%				
	C	O	Mg	N	Si, Ca
Untreated	80.5	15.3	-	2.0	2.2
10 min treated	43.7	42.3	2.9	3.6	7.5
30 min treated	42	43.6	5.6	3.0	5.8
60 min treated	34.7	45.4	10.1	2.8	7.0
90 min treated	28.8	41.9	19.0	3.5	6.7

The chemical bond analysis of Mung bean surface chemical elements was performed by XPS measurement. Figure 5 shows the C1s and Mg1s peaks fitting results of the initial and plasma-treated (30 and 60 min) Mung bean seeds. The analysis of the Mg1s peak showed two types of chemical bonds, MgO and $Mg(OH)_2$, with binding energies at 1304.5 and 1302.3 eV, respectively, which match well with the theoretical values [39]. The ratio between MgO and $Mg(OH)_2$ remains relatively stable (roughly 1:1) and Mg does not form any other bond despite a longer treatment time.

It should be mentioned that the analysis of the O1s peak confirmed the MgO and $Mg(OH)_2$ chemical bond formation with a similar area ratio as in the Mg1s spectra. Moreover, two additional chemical bonds were found (C-O and C=O). These chemical bonds were identified in detail by analyzing the C1s spectra. Nevertheless, any relevant information was found by analyzing the O1s spectra. Therefore, the fitting results of the O1s peak are not presented as a figure.

The results of carbon peak (C1s) deconvolution revealed that it consists of three components: C-C, C-O, and C=O at binding energies of 284.8, 286.1, and 288.3 eV, respectively. It is important to mention that the chemical bond C-C covers the majority of the C1s peak area by analyzing the initial seeds (76.0% of the area). However, this value decreased until 51.5% of the area after 90 min of the plasma treatment, while the total amount of C-O and C=O chemical bonds increased with each longer treatment time. It is known that carbon chains (C-C bond) are recognized as non-polar groups, which are responsible for surface hydrophobicity. Meanwhile, the higher amount of polar groups (C-O and C=O) increase the surface tension and induce its hydrophilicity [40]. These results revealed that the plasma treatment invoked a reduction of non-polar groups on the seeds surface and increased its energy. Hence, the surface energy increment determined the formation of strongly polar functional groups on the seeds surface [41,42]. The polar/non-polar groups ratio and chemical bonds concentrations of the C1s peak are presented in Table 2.

Table 2. Concentrations of polar/non-polar groups.

Samples	C1s, area %			Polar/Non-Polar Groups Ratio
	C-C	C-O	C=O	
Initial	76.0	17.7	6.3	0.32
10 min	71.1	24.3	4.6	0.40
30 min	60.8	25.7	13.5	0.64
60 min	55.0	37.6	7.4	0.82
90 min	51.5	30.6	17.9	0.94

The water contact angle measurement is a simple technique which is used for a qualitative evaluation of surface wettability. The WCA measurement was performed after each plasma treatment time, as well as for the initial Mung bean seeds. The significant reduction of WCA was observed by comparing initial and any other plasma-treated seeds. The WCA value for the initial seed was about 96°. This value decreased until approximately 38° after 1 min and 24° after 5 min. A further analysis revealed that the value of WCA remained relatively stable (about 15°) after 10 min of plasma treatment, while the ratio of polar/non-polar groups increased with a longer treatment time (Figure 6). Such an obvious reduction of WCA determined clear changes of the seeds surface. The surface conversion from hydrophobic to hydrophilic indicated the increment of free surface energy [40]. This, in turn, determined a better water absorption into the seed. G. de Groot et al. used the cold plasma treatment for cotton seed germination improvement [8]. They showed that the cold plasma treatment invoked the seeds surface hydrophilicity and improved water absorption. Moreover, they showed that the enlarged water absorption, compared to the initial seed, was directly related to a better seed germination. The similar germination tendencies were observed by other authors as well, when the plasma treatment encouraged the seeds surface hydrophilicity, which is directly related to a better germination process [10,43,44].

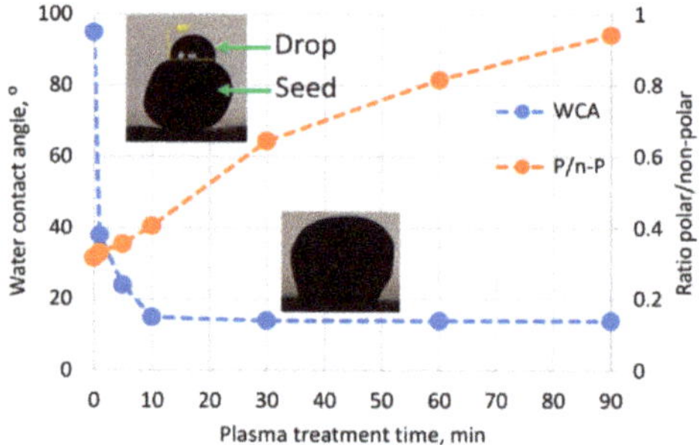

Figure 6. Water contact angle (WCA) and polar/non-polar groups ratio (P/n-P) dependence on the plasma treatment time.

The Mung bean germination experiments were performed with the seeds before and after the plasma treatment process. First, the initial seeds showed approximately 40% germination (Figure 7a). The increment of the germination process was observed after 10 and 30 min of the plasma treatment (50% and 70% germination, respectively). However, a longer treatment time determined a total reduction of the germination process (10% and 0% after 60 and 90 min of the plasma treatment, respectively). The results of the sprouts length average values with the standard deviation (SD) are shown in Figure 7a. The measurements revealed that the longest sprouts were observed after 30 min of the plasma treatment with the average values of 73.9 mm (SD—23.0). Meanwhile, the length of the initial and 10-min plasma-treated Mung bean sprouts were fixed at 71.3 mm (SD—22.8) and 65.2 mm (SD 11.3), respectively.

We speculate that 30 min of the plasma treatment leads to formation of the most suitable number of polar groups and, in turn, an enhanced water absorption, which is directly related to the seeds germination process. On the other hand, a longer treatment time (60 and 90 min) inhibits the germination process by initiating a too intensive polar groups formation (more than 0.6 of polar/non-polar groups ratio) and invokes an even higher water absorption through the surface sheath, as well as induces harmful stress in the seeds surface.

Magnesium plays an important role as a cationic macronutrient, which is a component of the chlorophyll molecule and participates in the activation of photosynthetic enzymes [24]. The deficiency of magnesium limits the photosynthesis process and leads to deprivation of carbohydrates in roots, seeds, and tubers [24,45].

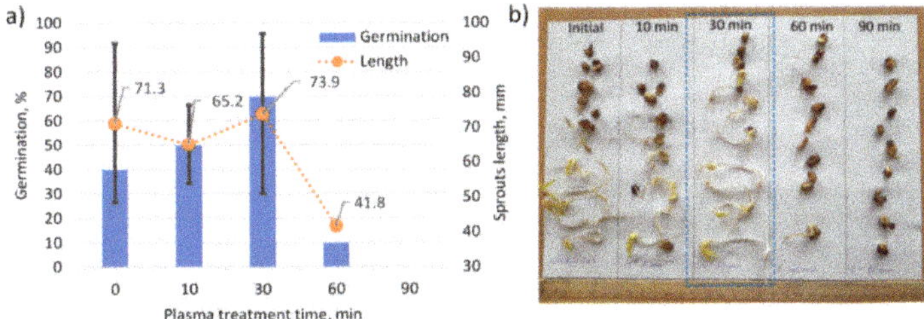

Figure 7. Mung bean germination: (**a**) Germination percentage with sprouts length averages and (**b**) real sprouts images of initial and plasma-treated seeds. Black vertical lines in (**a**)—errors presented as standard deviation.

Furthermore, the plasma treatment also helps improve the activity of numerous plant enzymes including seedling germination enzymes, which impact the overall metabolism of the plant and increase its growth [46]. On the other hand, the plasma treatment has a positive influence on the regulation of water uptake balance and physiological metabolic processes by modulating the activity of antioxidant enzymes [47,48].

The surface morphology views of Mung beans before and after the plasma treatment are shown in Figure 8. By comparing the initial and plasma-treated Mung beans surface, the results revealed that its structure remained relatively stable and does not form any obvious changes after 10 and 30 min of the plasma treatment (Figure 8b,c). Nevertheless, a clear evidence of surface morphology changes was observed after 60 and 90 min of the plasma treatment, which contains small and significant cracks (see the red dots in Figure 8d,e). It was assumed that the extensive plasma treatment could influence the appearance of cracks by composing surface stress. These cracks might act as a water uptake center itself and inhibit the seeds germination at higher treatment times, as seen in Figure 7.

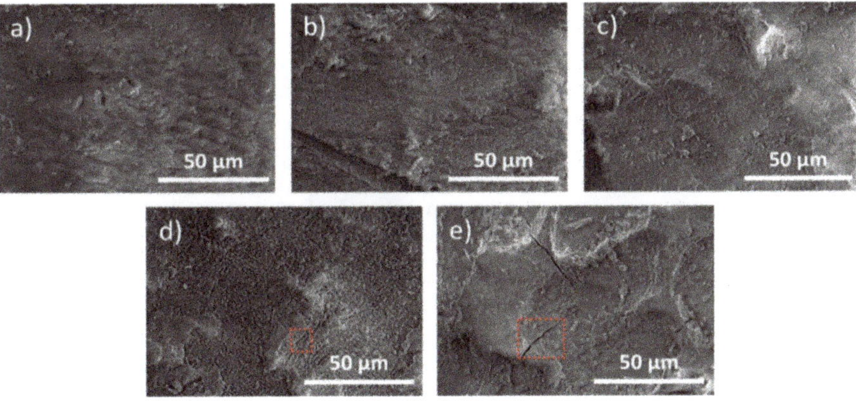

Figure 8. Morphology views of Mung beans: (**a**) Initial, (**b**) 10 min, (**c**) 30 min, (**d**) 60 min, (**e**) 90 min of the plasma treatment. Red dots—appeared cracks.

Moreover, as observed from the XPS elemental composition results, the amount of Mg increased with the longer treatment time (5.6 at% after 30 min and 19.0 at% after 90 min of the plasma treatment). The results of Mg elemental mapping after 30, 60, and 90 min of the plasma treatment are shown in Figure 9. The elemental mapping results showed that the Mg nanoparticles (red dots) distributed randomly on the seeds surface after 30 min of the plasma treatment and the highest part of the seeds surface remained uncovered by the Mg nanoparticles. The different results were observed after 60 and 90 min of the plasma treatment. The significant higher surface area was covered by Mg nanoparticles with a clear evidence of Mg clusters formation after 90 min of the plasma treatment. Presumably, the formation of enlarged Mg clusters might have a negative influence on the Mung bean seeds germination process, which act as a physical barrier. Nevertheless, further investigations of Mg, as well as the influence of other minerals on seeds germination and growth, are required.

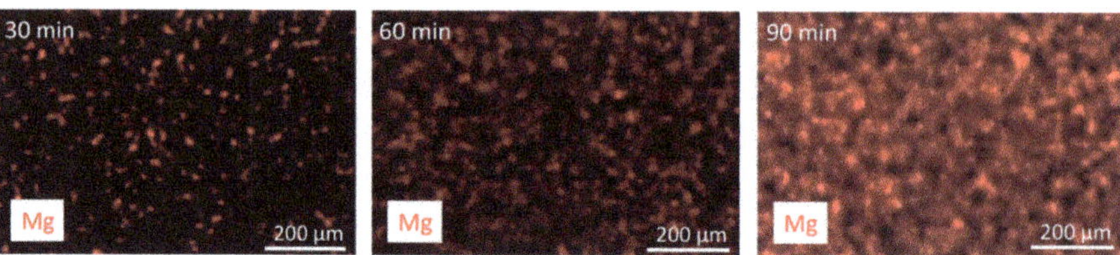

Figure 9. Elemental mapping results of Mg nanoparticles distribution on Mung bean seeds surface after 30, 60 and 90 min of the plasma treatment.

4. Conclusions

In this study, a new technique based on the combination of simultaneous non-thermal air plasma treatment with Mg nanoparticles deposition processes were applied to Mung bean seeds by enhancing their quality. This method revealed an original approach in order to increase the macronutrients level on the seeds, which could be controlled during the plasma treatment process. The simulation of the plasma treatment process showed that ions can penetrate into approximately 3 nm depth of Mung bean seeds, while Mg ions can penetrate into a 3 nm depth. The two processes known as plasma cleaning and free radical sites formation invoked composition changes on the seeds surface during the plasma treatment with the reduction of carbon and increment of oxygen components, as well as Mg nanoparticles deposition. This stimulates new chemical bonds formation on the seeds surface. The formation of polar chemical bonds (C-O and C=O) lead to the shift of surface characteristics from hydrophobic to hydrophilic and, in turn, improve water uptake. Meanwhile, the deposited Mg nanoparticles formed MgO and $Mg(OH)_2$ compounds, which can help in plant growth promotion. Finally, the experiments showed that almost a two times better germination (70% outcome with a 73.9 mm average sprouts length) was observed after 30 min of the plasma treatment compared to the initial Mung bean seeds (40% outcome with a 71.3 mm average sprouts length).

Author Contributions: Writing—original draft preparation, writing—review and editing, S.V. (Sarunas Varnagiris) and M.U.; investigation, formal analysis, and methodology, S.V. (Simona Vilimaite), I.M. and S.T.; conceptualization and supervision, D.M. All authors have read and agreed to the published version of the manuscript.

Funding: This research received no external funding.

Conflicts of Interest: The authors declare no conflict of interest.

References

1. Zhou, R.; Li, J.; Zhou, R.; Zhang, X.; Yang, S. Atmospheric-pressure plasma treated water for seed germination and seedling growth of mung bean and its sterilization effect on mung bean sprouts. *Innov. Food Sci. Emerg. Technol.* **2019**, *53*, 36–44. [CrossRef]
2. Xiang, Q.; Liu, X.; Liu, S.; Ma, Y.; Xu, C.; Bai, Y. Effect of plasma-activated water on microbial quality and physicochemical characteristics of mung bean sprouts. *Innov. Food Sci. Emerg. Technol.* **2019**, *52*, 49–56. [CrossRef]
3. Šerá, B.; Šerý, M. Non-thermal plasma treatment as a new biotechnology in relation to seeds, dry fruits, and grains. *Plasma Sci. Technol.* **2018**, *20*, 044012. [CrossRef]
4. Sivachandiran, L.; Khacef, A. Enhanced seed germination and plant growth by atmospheric pressure cold air plasma: Combined effect of seed and water treatment. *RSC Adv.* **2017**, *7*, 1822–1832. [CrossRef]
5. Bibwe, B.; Kannaujia, P.K.; Mahawar, M.K.; Aradwad, P. Cold plasma - Improving seed germination and seedling vigour. *Agric. Food* **2019**, *1*, 1–4.

6. Morison, J.I.L.; Baker, N.R.; Mullineaux, P.M.; Davies, W.J.; Morison, J.I.L.; Baker, N.R.; Mullineaux, P.M.; Davies, W.J. Improving water use in crop production. *Philos. Trans. R. Soc. B Biol. Sci.* **2008**, *363*, 639–658. [CrossRef]
7. Pérez-Pizá, M.C.; Prevosto, L.; Grijalba, P.E.; Zilli, C.G.; Cejas, E.; Mancinelli, B.; Balestrasse, K.B. Improvement of growth and yield of soybean plants through the application of non-thermal plasmas to seeds with different health status. *Heliyon* **2019**, *5*, e01495. [CrossRef]
8. de Groot, G.J.J.B.; Hundt, A.; Murphy, A.B.; Bange, M.P.; Mai-Prochnow, A. Cold plasma treatment for cotton seed germination improvement. *Sci. Rep.* **2018**, *8*, 1–10. [CrossRef]
9. Huang, W.Z.; Hsiao, A.I. Factors affecting seed dormancy and germination of Paspalum distichum. *Weed Res.* **1987**, *27*, 405–415. [CrossRef]
10. Molina, R.; López-Santos, C.; Gómez-Ramírez, A.; Vílchez, A.; Espinós, J.P.; González-Elipe, A.R. Influence of irrigation conditions in the germination of plasma treated Nasturtium seeds. *Sci. Rep.* **2018**, *8*, 1–11. [CrossRef]
11. Lotfy, K.; Al-Harbi, N.A.; Abd El-Raheem, H. Cold Atmospheric Pressure Nitrogen Plasma Jet for Enhancement Germination of Wheat Seeds. *Plasma Chem. Plasma Process.* **2019**, *39*, 897–912. [CrossRef]
12. Zhang, B.; Li, R.; Yan, J. Study on activation and improvement of crop seeds by the application of plasma treating seeds equipment. *Arch. Biochem. Biophys.* **2018**, *655*, 37–42. [CrossRef] [PubMed]
13. Adamovich, I.; Baalrud, S.D.; Bogaerts, A.; Bruggeman, P.J.; Cappelli, M.; Colombo, V.; Czarnetzki, U.; Ebert, U.; Eden, J.G.; Favia, P.; et al. The 2017 Plasma Roadmap: Low temperature plasma science and technology. *J. Phys. D Appl. Phys.* **2017**, *50*, 323001. [CrossRef]
14. Khamsen, N.; Onwimol, D.; Teerakawanich, N.; Dechanupaprittha, S.; Kanokbannakorn, W.; Hongesombut, K.; Srisonphan, S. Rice (*Oryza sativa* L.) Seed Sterilization and Germination Enhancement via Atmospheric Hybrid Nonthermal Discharge Plasma. *ACS Appl. Mater. Interfaces* **2016**, *8*, 19268–19275. [CrossRef]
15. Los, A.; Ziuzina, D.; Boehm, D.; Cullen, P.J.; Bourke, P. Investigation of mechanisms involved in germination enhancement of wheat (*Triticum aestivum*) by cold plasma: Effects on seed surface chemistry and characteristics. *Plasma Process. Polym.* **2019**, *16*, 1–12. [CrossRef]
16. Dobrin, D.; Magureanu, M.; Mandache, N.B.; Ionita, M.D. The effect of non-thermal plasma treatment on wheat germination and early growth. *Innov. Food Sci. Emerg. Technol.* **2015**, *29*, 255–260. [CrossRef]
17. Nalwa, C.; Thakur, A.K. Seed quality enhancement through plasma treatment: A review. *Indian J. Ecol.* **2018**, *45*, 814–821.
18. Ito, M.; Hori, M. Current status and future prospects of agricultural applications using atmospheric-pressure plasma technologies. *Plasma Process. Polym.* **2017**, *15*, 1700073. [CrossRef]
19. Măgureanu, M.; Daniela, R.S.; Mihai, D. Stimulation of the Germination and Early Growth of Tomato Seeds by Non-thermal Plasma. *Plasma Chem. Plasma Process.* **2018**, *38*, 989–1001. [CrossRef]
20. Ahn, C.; Gill, J.; Ruzic, D.N. Growth of Plasma-Treated Corn Seeds under Realistic Conditions. *Sci. Rep.* **2019**, *9*, 1–7. [CrossRef]
21. Scholtz, V.; Šerá, B.; Khun, J.; Šerý, M.; Julák, J. Effects of Nonthermal Plasma on Wheat Grains and Products. *J. Food Qual.* **2019**, *2019*, 7917825. [CrossRef]
22. Sadhu, S.; Thirumdas, R.; Deshmukh, R.R.; Annapure, U.S. Influence of cold plasma on the enzymatic activity in germinating mung beans (*Vigna radiate*). *LWT Food Sci. Technol.* **2017**, *78*, 97–104. [CrossRef]
23. Bogdanov, T.; Tsonev, I.; Marinova, P.; Benova, E.; Rusanov, K.; Rusanova, M.; Atanassov, I.; Kozáková, Z.; Krčma, F. Microwave plasma torch generated in argon for small berries surface treatment. *Appl. Sci.* **2018**, *8*, 1870. [CrossRef]
24. Ceylan, Y.; Kutman, U.B.; Mengutay, M.; Cakmak, I. Magnesium applications to growth medium and foliage affect the starch distribution, increase the grain size and improve the seed germination in wheat. *Plant. Soil* **2016**, *406*, 145–156. [CrossRef]
25. Cai, L.; Liu, M.; Liu, Z.; Yang, H.; Sun, X.; Chen, J.; Xiang, S.; Ding, W. MgoNPs can boost plant growth: Evidence from increased seedling growth, morpho-physiological activities, and Mg uptake in Tobacco (*Nicotiana tabacum* L.). *Molecules* **2018**, *23*, 3375. [CrossRef]
26. Shinde, S.; Paralikar, P.; Ingle, A.P.; Rai, M. Promotion of seed germination and seedling growth of Zea mays by magnesium hydroxide nanoparticles synthesized by the filtrate from Aspergillus niger. *Arab. J. Chem.* **2020**, *13*, 3172–3182. [CrossRef]

27. Vijai Anand, K.; Anugraga, A.R.; Kannan, M.; Singaravelu, G.; Govindaraju, K. Bio-engineered magnesium oxide nanoparticles as nano-priming agent for enhancing seed germination and seedling vigour of green gram (*Vigna radiata* L.). *Mater. Lett.* **2020**, *271*, 127792. [CrossRef]
28. Zhang, B.; Cakmak, I.; Feng, J.; Yu, C.; Chen, X.; Xie, D.; Wu, L.; Song, Z.; Cao, J.; He, Y. Magnesium Deficiency Reduced the Yield and Seed Germination in Wax Gourd by Affecting the Carbohydrate Translocation. *Front. Plant. Sci.* **2020**, *11*, 1–10. [CrossRef]
29. Yuan, M.; Zhang, X.; Saeedi, A.M.A.; Cheng, W.; Guo, C.; Liao, B.; Zhang, X.; Ying, M.; Gehring, G.A. Study of the radiation damage caused by ion implantation in ZnO and its relation to magnetism. *Nucl. Instrum. Methods Phys. Res. Sect. B Beam Interact. Mater. Atoms* **2019**, *455*, 7–12. [CrossRef]
30. Medvids, A.; Varnagiris, S.; Letko, E.; Milcius, D.; Grase, L.; Gaidukovs, S.; Mychko, A.; Pludons, A.; Onufrijevs, P.; Mimura, H. Phase transformation from rutile to anatase with oxygen ion dose in the TiO_2 layer formed on a Ti substrate. *Mater. Sci. Semicond. Process.* **2020**, *106*, 104776. [CrossRef]
31. Greczynski, G.; Hultman, L. Reliable determination of chemical state in x-ray photoelectron spectroscopy based on sample-work-function referencing to adventitious carbon: Resolving the myth of apparent constant binding energy of the C 1s peak. *Appl. Surf. Sci.* **2018**, *451*, 99–103. [CrossRef]
32. Hertwig, C.; Meneses, N.; Mathys, A. Cold atmospheric pressure plasma and low energy electron beam as alternative nonthermal decontamination technologies for dry food surfaces: A review. *Trends Food Sci. Technol.* **2018**, *77*, 131–142. [CrossRef]
33. Dubinov, A.E.; Lazarenko, E.M.; Selemir, V.D. Effect of Glow Discherge Air Plasma on Grain Corps Seef. *IEEE Trans. Plasma Sci.* **2000**, *28*, 180–183. [CrossRef]
34. Conrad, J.R. *Handbook of Plasma Immersion Ion. Implantation and Deposition*; Anders, A., Ed.; John Wiley & Sons: New York, NY, USA, 2000.
35. Rifna, E.J.; Ramanan, K.R.; Mahendran, R. Emerging technology applications for improving seed germination. *Trends Food Sci. Technol.* **2019**, *86*, 95–108. [CrossRef]
36. Bourke, P.; Ziuzina, D.; Boehm, D.; Cullen, P.J.; Keener, K. The Potential of Cold Plasma for Safe and Sustainable Food Production. *Trends Biotechnol.* **2018**, *36*, 615–626. [CrossRef] [PubMed]
37. Butscher, D.; Waskow, A.; von Rohr, P.R. Chapter 6—Disinfection of granular food products using cold plasma. In *Advances in Cold Plasma Applications for Food Safety and Preservation*; Bermudez-Aguirre, D., Ed.; Academic Press: Cambridge, MA, USA, 2020; pp. 185–228, ISBN 978-0-12-814921-8.
38. Dhayal, M.; Alexander, M.R.; Bradley, J.W. The surface chemistry resulting from low-pressure plasma treatment of polystyrene: The effect of residual vessel bound oxygen. *Appl. Surf. Sci.* **2006**, *252*, 7957–7963. [CrossRef]
39. NIST. NIST X-ray photoelectron spectroscopy database. In *NIST Standard Reference Database 20*; Version 4.1. Available online: https://srdata.nist.gov/xps/ (accessed on 29 November 2020).
40. Long, J.; Zhong, M.; Zhang, H.; Fan, P. Superhydrophilicity to superhydrophobicity transition of picosecond laser microstructured aluminum in ambient air. *J. Colloid Interface Sci.* **2015**, *441*, 1–9. [CrossRef] [PubMed]
41. Urbonavicius, M.; Varnagiris, S.; Milcius, D. Generation of hydrogen through the reaction between plasma-modified aluminum and water. *Energy Technol.* **2017**, *5*, 2300–2308. [CrossRef]
42. Azimi, G.; Dhiman, R.; Kwon, H.; Paxson, A.T.; Varanasi, K.K. Hydrophobicity of rare-earth oxide ceramics. *Nat. Mater.* **2013**, *12*, 315–320. [CrossRef]
43. Nantapan, J.; Sarapirom, S.; Janpong, K. The effects of atmospheric plasma jet treatment to the germination and enhancement growth of sunflower seeds. *J. Phys. Conf. Ser.* **2019**, *1380*, 012131. [CrossRef]
44. Bormashenko, E.; Grynyov, R.; Bormashenko, Y.; Drori, E. Cold radiofrequency plasma treatment modifies wettability and germination speed of plant seeds. *Sci. Rep.* **2012**, *2*, 3–10. [CrossRef] [PubMed]
45. Naseeruddin, R.; Sumathi, V.; Prasad, T.N.V.K.V.; Sudhakar, P.; Chandrika, V.; Reddy, B.R. Unprecedented Synergistic Effects of Nanoscale Nutrients on Growth, Productivity of Sweet Sorghum [*Sorghum bicolor* (L.) Moench], and Nutrient Biofortification. *J. Agric. Food Chem.* **2018**, *66*, 1075–1084. [CrossRef] [PubMed]
46. Li, L.; Jiang, J.; Li, J.; Shen, M.; He, X.; Shao, H.; Dong, Y. Effects of cold plasma treatment on seed germination and seedling growth of soybean. *Sci. Rep.* **2014**, *4*, 5859.
47. Ling, L.; Jiangang, L.; Minchong, S.; Chunlei, Z.; Yuanhua, D. Cold plasma treatment enhances oilseed rape seed germination under drought stress. *Sci. Rep.* **2015**, *5*, 1–10. [CrossRef] [PubMed]

48. Henselová, M.; Slováková, Ľ.; Martinka, M.; Zahoranová, A. Growth, anatomy and enzyme activity changes in maize roots induced by treatment of seeds with low-temperature plasma. *Biologia* **2012**, *67*, 490–497. [CrossRef]

Publisher's Note: MDPI stays neutral with regard to jurisdictional claims in published maps and institutional affiliations.

Review

Plasma Agriculture from Laboratory to Farm: A Review

Pankaj Attri [1,*], Kenji Ishikawa [1], Takamasa Okumura [1,2,3], Kazunori Koga [1,2,3] and Masaharu Shiratani [1,2,*]

1. Center of Plasma Nano-interface Engineering, Kyushu University, Fukuoka 819-0395, Japan; ishikawa@plasma.ed.kyushu-u.ac.jp (K.I.); t.okumura@plasma.ed.kyushu-u.ac.jp (T.O.); koga@ed.kyushu-u.ac.jp (K.K.)
2. Department of Electronics, Kyushu University, Fukuoka 819-0395, Japan
3. Center for Novel Science Initiatives, National Institute of Natural Science, Tokyo 105-0001, Japan

* Correspondence: chem.pankaj@gmail.com (P.A.); siratani@ed.kyushu-u.ac.jp (M.S.); Tel.: +81-92-802-3723 (P.A. & M.S.)

Received: 30 July 2020; Accepted: 14 August 2020; Published: 17 August 2020

Abstract: In recent years, non-thermal plasma (NTP) application in agriculture is rapidly increasing. Many published articles and reviews in the literature are focus on the post-harvest use of plasma in agriculture. However, the pre-harvest application of plasma still in its early stage. Therefore, in this review, we covered the effect of NTP and plasma-treated water (PTW) on seed germination and growth enhancement. Further, we will discuss the change in biochemical analysis, e.g., the variation in phytohormones, phytochemicals, and antioxidant levels of seeds after treatment with NTP and PTW. Lastly, we will address the possibility of using plasma in the actual agriculture field and prospects of this technology.

Keywords: non-thermal plasma; plasma agriculture; seed germination; growth enhancement

---

1. Introduction

Plants regularly undergo a multitude of stresses, e.g., scarcity of water, waterlogging, toxicity, high salinity, and extreme temperatures. These stresses result in less yield of crops. Countries such as India, Australia, China, USA, South America, Central Asia, and Africa are significant producers of food crops and are facing droughts at regular intervals. To enhance seed germination and growth under the changing environment, techniques such as chemical, physical, and biological treatments are developing [1–4]. However, in the framework of seed technology, the physical invigoration treatments in seeds can result in the change of seed morphology, gene expression, and protein level [5]. These physical changes can result in increased germination and growth enhancement. The physical methods for pre-sowing seed treatments are magnetic fields, electromagnetic waves, ionizing radiations, ultrasounds, non-thermal plasma, etc., as shown in Figure 1. In this review, we will discuss the role of non-thermal plasma (NTP) in stimulating germination and growth in plant seeds.

NTP has received considerable attention in recent years due to its increasing applications in medicine, sterilization, agriculture, etc., as displayed in Figure 2 [6–26]. NTP discharges generate the reactive charged species such as electrons and ions, and neutral species, and emit ultraviolet radiation, and electric fields. The plasma generated reactive oxygen and nitrogen species (RONS) and change in solution properties pH, electrical conductivity, and oxidation-reduction potential. These solutions affect the rate of seed germination, enhancement in plant growth, and well as an increase in agricultural yields. NTP applications in agriculture possess advantages over conventional treatments such as short treatment time, easily accessible, and low temperature during operations. Moreover, NTP can be used to treat seeds and crops without damaging them. The use of different feeding gases can alter the plasma chemistry that leads to an increase in variation of seed coating technology in comparison to traditional methods [8].

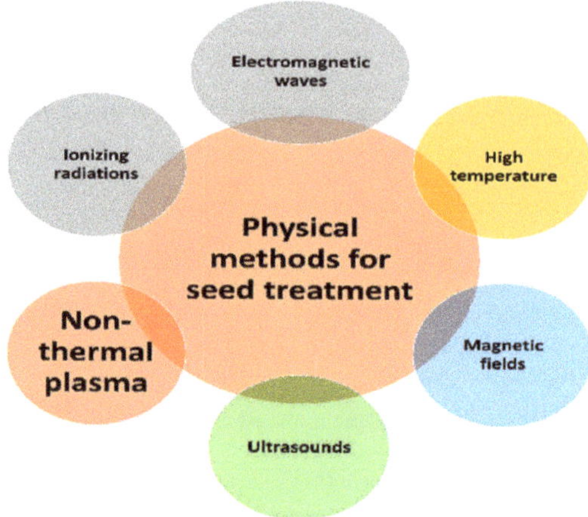

Figure 1. Schematic representation of various physical methods used for seed treatment.

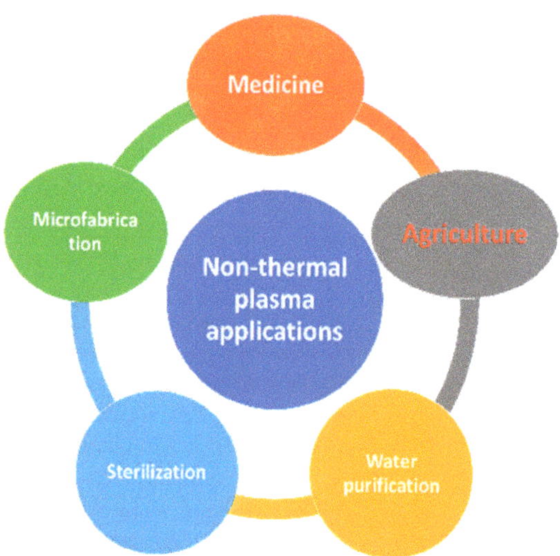

Figure 2. Schematic representation of non–thermal plasma (NTP) applications.

Generally, plasma applications in agriculture classified as preharvest and post-harvest. The use of NTP in post-harvest processes such as food preservation and food processing are discussed briefly in previously published reviews, as demonstrated in Figure 3 [27–32] On the other hand, limited reviews are available for the NTP use in preharvest [33–36]. NTP technology is used in preharvest at different levels like sterilization of seeds, improving seed germination, reducing pathogen invasion in soil, etc. Although in this review, we will focus on the role of NTP generated at variable pressure ranges on seed germination and seedling growth. In addition, we will also discuss the possible impact of plasma-treated water (PTW) (also known as the plasma-activated water (PAW)) on seed germination and plant growth. At last, we will discuss the probable mechanism of NTP and PTW treatment in plasma agriculture and the prospects of this technology in the real scenario.

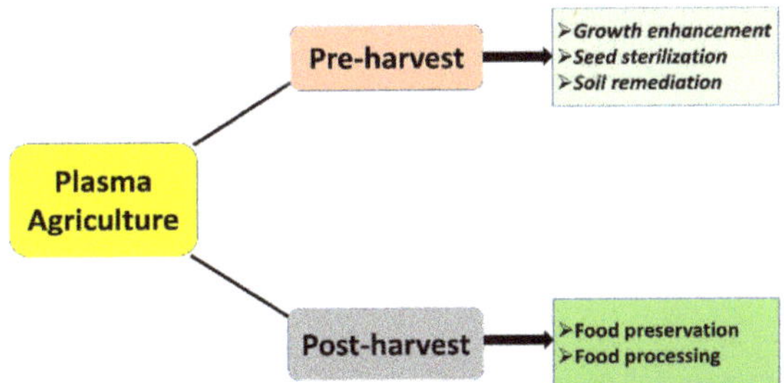

Figure 3. Role of plasma in preharvest and post-harvest processes.

This review aims to discuss the current status of plasma treatment in the pre-harvest stage and its possible mechanism. Additionally, explore the possibilities of using plasma in the actual agriculture field.

2. Effect of NTP at Low/Medium ($\approx 6.7 \times 10^{-2}$ to 53,328 Pa) Pressure on Seed Germination and Growth Enhancement

The NTP treatment applied at low or medium pressure, to treat various seed for the different time intervals, in the presence of different types of feeding gases as described in Table 1.

Table 1. Enhanced seed germination and growth due to low/medium/atmospheric pressure plasma and plasma-treated water treatment, as reported in the literature.

Seeds	Plasma Type	Results	Reference
Radish sprouts	Low-pressure plasma (100 Pa)	Radish sprouts growth increases with O_2 plasma treatment, while no effect was observed for seed germination. In contrast, no effect on the average length of sprouts for N_2 Low-pressure plasma.	[37]
Radish sprouts	Low-pressure plasma (40 Pa)	Growth enhancement.	[38]
Radish sprouts	Scalar dielectric-barrier discharge (DBD) plasma	Growth enhancement.	[39]
Radish sprouts	Scalar DBD plasma	The average seedling length was 250% longer than the control samples.	[40]
Radish sprouts	Scalar DBD plasma	Enhanced plant growth for O_2, Air and NO (10%) + N_2 feeding gases plasma. While no significant growth enhancement for He, N_2, and Ar gases plasma.	[41]
Radish sprouts	Plasma jet	The total mass and average lengths of radish sprouts increased.	[42]
Radish	Surface discharge plasma	No effect on the germination dynamics but the length of root and sprout increased.	[43]
Chili pepper	Plasma jet	Enhanced seed germination.	[44]
Arabidopsis thaliana	Low-pressure plasma (20–80 Pa)	Lengths of the leaves and stems of Arabidopsis increased $\approx$1.5 times over the control.	[45]
Arabidopsis thaliana	Scalar DBD plasma	Enhanced growth, shorter harvest, and increased total weight.	[46]
Wheat	Medium pressure glow discharge plasma ($\approx$1333 Pa)	Increased growth activity and dry matter accumulation.	[47]
Wheat	Low-pressure plasma	Plasma treatment increased the grain and spike yield.	[48]
Wheat	Low-pressure plasma	Enhanced seed germination rate.	[49]
Wheat	Low-pressure plasma (150 Pa)	Improved germination potential, germination rate.	[50]
Wheat	Surface discharge plasma	Little effect on the germination rate while a substantial impact on growth parameters.	[51]
Wheat	DBD plasma	Improved the germination and seedling growth.	[52]
Wheat	DBD plasma	The germination rate, germination potential, germination index, and vigor index increased after plasma treatment.	[53]

Table 1. *Cont.*

Seeds	Plasma Type	Results	Reference
Wheat	DBD plasma	The germination rate, germination potential, root length, and shoot length of the wheat seedlings increased.	[54]
Wheat	Plasma jet	Increased dry weight after plasma treatment.	[55]
Wheat	Low-pressure plasma (140 Pa)	Germination acceleration was inhibited on first day after plasma treatment.	[6]
Sunflower	Scalar DBD plasma	Adverse effects on germination kinetics.	[56]
Sunflower	Streamer like plasma	Growth enhancement and increased dry weight.	[57]
Sunflower	DBD plasma	The distribution of sprouts length and the dry weight increased after plasma treatment.	[58]
Soybean	Low-pressure plasma (150 Pa)	Germination and vigor indices significantly increased after plasma treatment.	[59]
Soybean	DBD plasma	Total fresh weight increased by 1.2-fold for DBD plasma	[60]
Pea	Diffuse coplanar surface barrier discharge (DCSBD) plasma	Increased in germination percentage and growth parameters.	[61]
Pea and Zucchini	FSG plasma (a semi-automatic device) system.	Germination of Pea and Zucchini increased after plasma treatment.	[62]
Mung bean	Microplasma array plasma.	Germination index increased for Air and O_2 plasma, and no significant difference observed for He or N_2 plasma compared to control.	[63]
Beans	Low-pressure plasma (6.7×10^{-2} Pa)	The final germination percentage of seeds was not affected by plasma treatment. However, the rate of germination was improved for the plasma-treated samples.	[64]
Artichoke	Low-pressure plasma (1.8 Pa)	Improved the germination rate and seedling growth.	[65]
Ajwain	Low-pressure plasma (9.9 Pa)	Improved seed germination percentage and germination index.	[66]
Poppy	Plasonic AR-550-M	Enhanced seed germination	[67]
Oilseed rape	Low-pressure plasma (150 Pa)	Improved germination rate and seedling growth.	[68]
Hemp	Gliding arc and downstream microwave devices (low-pressure, 140 Pa)	Gliding arc treatment increased the length of seedlings, seedling accretion, and weight of seedling, while downstream microwave plasma treatment had an inhibiting effect.	[69]
Garlic seed bulbs	Low-pressure plasma (15–60 Pa)	Increased dried bulb mass after plasma treatment.	[70]
Sweet basil	Low-pressure plasma (40 Pa)	Increased germination and seedling vigor after plasma treatment.	[71]
Black gram	Medium pressure DBD plasma (≈53,328 Pa)	Enhanced seed germination rate and seedling growth.	[72]
Tomato	Coaxial DBD reactor plasma	The root-to-shoot ratio (R/S) ratio increased significantly for plasma-treated samples.	[73]
Pumpkin	Plasma jet	Plasma jets accelerated the germination of pumpkin seeds.	[74]
Brassica napus	DBD plasma	No significant difference in seed germination.	[75]
Andrographis paniculata	DBD plasma	Increased seed germination.	[76]
Fenugreek	Plasma jet.	Enhanced seed germination rate.	[77]
Mulungu	Plasma jet	Enhanced seed germination rate.	[78]
Hybanthus calceolaria	Plasma jet	Enhanced seed germination rate.	[79]
Nasturtium	DBD plasma	Enhanced seed germination for short plasma treatment.	[80]
Thuringian Mallow	GlidArc reactor	Enhanced seed germination.	[81]
Cucumber and Pepper	DCSBD plasma	Improved germination observed for both seeds.	[82]
Spinach	High voltage nanosecond pulsed plasma and micro DBD plasma.	Germination and dry weight of seedlings increased after both plasma treatment.	[83]
Barley	Surface DBD plasma	Accelerated the early growth of sprouts and enhance bioactive phytochemicals in the sprouts.	[84]
Barley	Low-pressure plasma (≈26 Pa)	No effect of plasma treatment.	[7]

Table 1. *Cont.*

Seeds	Plasma Type	Results	Reference
Oat	Low-pressure plasma (≈13 Pa)	Quantity of germination seeds increased by 27% after plasma treatment than control on 5th day.	[7]
Oat	Low-pressure plasma (140 Pa)	No significant difference in rate of germination.	[6]
Black Pine	DCSBD plasma	The germination index increased for short treatment time.	[85]
Basil	Volume DBD plasma	Increased overall germination rate.	[86]
Black Gram	PTW	Increased cumulative germination and vigor index.	[87]
Radish sprout	PTW	Increased growth of sprouts.	[88]
Radish	PTW	Enhanced seeds germination rate and the seedling growth.	[89]
Soybeans	PTW	Enhanced seeds germination.	[90]
Mung bean	PTW	No significant difference in growth rate.	[91]
Zinnia annual	PTW	Increased germinability and growth of flowers of Zinnia annual.	[92]
Chinese Cabbage	PTW	Increased dried weight of the plant.	[93]
Lentils	PTW	Enhanced seeds germination as compared with commercial fertilizer.	[94]
Tomato	PTW	Enhanced shoot and root length.	[95]
Rapeseed	PTW	Significant improvement in germination rate and seedling vigor.	[96]

Low/medium-pressure plasma used to treat the seeds of radish sprouts, wheat, ajwain, black gram, poppy, oilseed rape, garlic, sweet basil, and bean. Radish Sprouts (*Raphanus sativus*) seeds were treated by low-pressure radiofrequency (RF) plasma at 100 Pa pressure with O_2 and N_2 as feed gases. The discharge power and frequency were 50 W and 13.56 MHz, respectively, see Figure 4a [37].

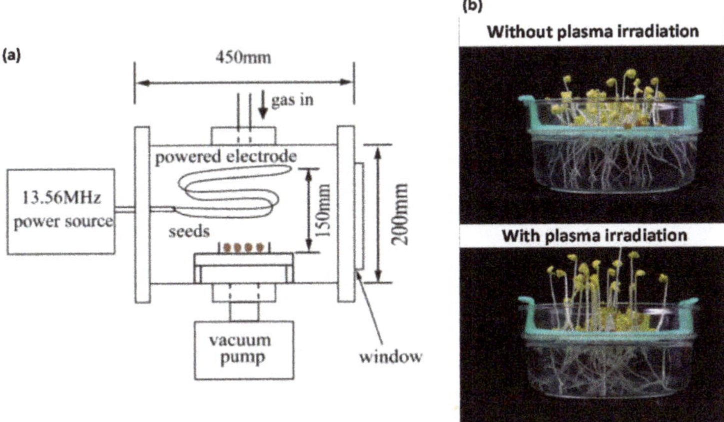

Figure 4. (**a**) Schematic diagram of the low-temperature plasma reactor and (**b**) Radish sprouts cultivated with and without O_2 plasma irradiation. Reproduced from reference [37], Copyright (2012) The Japan Society of Applied Physics.

No effect on the seed germination was reported for O_2 and N_2 low-pressure plasma treatment. However, the average length of sprouts was 60% higher for O_2 low-pressure plasma as compared to the control sample, as indicated in Figure 4b. In contrast, no change was observed for N_2 low-pressure plasma [37]. Hayashi et al. studied the effect of inductively coupled RF plasma on the radish sprouts seeds. RF plasma worked at 13.56 MHz frequency, 40 Pa pressure in the presence of ambient air for 20 min, with an input power of 50 W. Enhancement in radish sprouts growth observed in their study [38].

In another study, *Arabidopsis thaliana* and *Raphanus sativus* seeds were treated with O_2 low-pressure plasma with RF power at power 60 W, 20–80 Pa pressure, and 13.56 MHz frequency. The length of *Raphanus sativus* varied with changes in the pressure, although at pressure 20 Pa maximum length was obtained. Whereas, plasma treatment on *Arabidopsis thaliana* seeds results in increased length of stems by 1.5 times and area of leaves by 2 times as compared with control (without plasma treatment) [45].

Wheat is an essential strategic crop; therefore, many researchers used NTP to treat the wheat seeds. The wheat seeds (*Triticum aestivum*) treated by glow discharge plasma with a mixture of air/O_2 gases at ≈1333 Pa pressure and 3–5 kHz frequency for 3–9 min. They observed that 6 min treatment of glow discharge plasma could result in 95–100% seed germination and a 20% increase in wheat yield [47]. In a very recent study, RF plasma reactor operated at 13.56 MHz frequency in the air for 180 s used to treat the same wheat seeds (*Triticum* spp.). They found that after plasma treatment, the grain and spike yield was enhanced to 58 and 75%, respectively, compared to control in the presence of haze stress [48]. Iqbal et al. treated the wheat (Galaxy-2013) seeds with Ar low-pressure plasma at variable voltages (600–850 V). They observed that the germination rate was 57–60% higher with changing the voltage from 600 to 850 V as compared to control [49]. Another group used the same wheat seeds (*Triticum* spp.) and treated with He plasma for 15 s at 150 Pa pressure and 3×10^9 MHz frequency with 60–100 W variable power. The plasma treatment at 80 W showed 6 and 6.7%, improvement in seed germination potential and germination rate, respectively, as compared to control. Additionally, the plant height, root length, and fresh weight increased to 20.3, 9, and 21.8%, respectively, at the seedling stage. At the same time, the wheat yield was increased to 5.89% with respect to control [50]. Šerá et al. showed that wheat germination acceleration was inhibited on the first day after treatment with Plasonic AR-550-M at power 500 W and pressure 140 Pa, whereas no significant effect observed for Oat caryopses treatment [6]. On the other hand, Dubinov et al. revealed 27% increase in the quantity of germination of Oat seeds treated with glow discharge plasma at pressure ≈13 Pa with respect to reference seeds [7].

Bormashenko et al. used the inductive air plasma discharge to treat beans (*Phaseolus vulgaris* L.) seeds for 2 min at 10 MHz frequency, 6.7×10^{-2} Pa pressure, and 20 W power. No significant change in germination percentage was observed between plasma-treated sample and control, although speed of germination was faster for plasma-treated samples than control [64]. In a sperate study, a commercial computer-controlled plasma device (HD-2N) treated the soybean (*Glycine max)* seeds. HD-2N plasma device works at a frequency of 13.56 MHz and pressure of 150 Pa with variable powers from 60 to 120 W. The improvement in the seed germination and seedling growth obtained at 80 W power. Shoot length, shoot dry weight, root length, and root dry weight increased by 13.77, 21.95, 21.42, and 27.51%, respectively, after plasma treatment with respect to control [59].

Hosseini et al. treated the artichoke seeds (*Cynara cardunculus* var. *scolymus* L.) with capacitively coupled RF plasma at a working pressure of 1.8 Pa with the power of 10 W. Authors showed that length of root increased by 28.5 and 50% after 10 and 15 min plasma treatment, respectively. The dry weight of roots was increased by 13 and 53% after 10 and 15 min plasma treatment, respectively [65]. Gholami et al. treated the ajwain seeds by RF capacitively coupled plasma at a frequency of 13.56 MHz and pressure of 9.9 Pa for 2 min at variable power. Ajwain seeds treated at 50 W power plasma showed 11.1, and 1.22% increase in germination percentage and germination index, respectively. At the same condition, root length was increased by 34% as compared to control. However, root length increments were 2 and 10% at powers of 80 and 100 W, respectively. The authors concluded that the germination percentage and germination index values decreased when power was more than 50 W [66].

Poppy seeds (*Papaver somniferum*) treated with 2.45 GHz microwave power source with a magnetron input of 500 W (Plasonic AR-550-M) at different time intervals. The seed germination rate on the fifth day was a maximum of 104 and 102% for 3 and 5 min plasma treatment, respectively, as compared to reference samples [67]. Ling et al. showed that oilseed rape (*Brassica napus* L.) seeds treated for 15 s with He-plasma at 13.56 MHz frequency, 100 W power, and 150 Pa pressure results in the improved germination rate of Zhongshuang 7 and Zhongshuang 11 by 6.25 and 4.44%, respectively [68]. Sera et al. investigated the effect of low-pressure plasma on Hemp (*Cannabis sativa* L.) seeds. They used

2.45 GHz microwave power source with a magnetron input of 500 W at a pressure of 140 Pa in the presence of O_2 and Ar gases. The authors observed the inhibitory effect of plasma treatment on all tested hemp cultivars [69].

Recently, garlic seed bulbs (*Ptujski spomladanski*) treated with O_2 low-pressure RF plasma at 15–60 Pa pressure for 60 s. The authors noticed increased dried bulb mass by 11% at 15 Pa pressure. Additionally, treatment at 30 and 45 Pa pressure exhibit little increase in dry bulb mass. Further, an increase in pressure (60 Pa) results in decreased dried bulb mass [70]. Another recent study by Singh et al. showed the increased in germination percent of sweet basil (*Ocimum basilicum* L.) seeds when treated with RF plasma at 13.56 MHz frequency, 40 Pa pressure with variable power 30–270 W in the mixture of O_2 (80%) and Ar (20%) gases. The germination percentage increased by 16.3 and 20.5% than control at power 90 and 150 W, respectively [71].

In a very recent study, DBD plasma working at a pressure of ≈53,328 Pa with 45 W power at the applied voltage of 5 kV and frequency of 4.5 kHz used to treat black gram (*Vigna mungo*) seeds. After plasma treatment, the rate of seed germination and seedling growth was increased by 13.67 and 37.13%, respectively, as compared to control [72].

3. Effect of NTP at Atmospheric Pressure on Seed Germination and Growth Enhancement

Recently reported work reveals the increasing use of atmospheric pressure non-thermal plasma (AP-NTP) than low–temperature plasma to treat the seeds. This increase was due to the difference in the treatment cost of both devices as well as the user-friendly operation of atmospheric pressure plasma devices (easily operated). The various seeds, such as radish sprouts, wheat, sunflower, pea, bean, maize, rice, pumpkin, cucumber, pepper, barley, spinach, basil, black pine, etc., were treated by AP-NTP. It was showed that scalar dielectric-barrier discharge (DBD) treated the *Raphanus sativus*, *Oryza sativa*, *Arabidopsis Thaliana*, *Plumeria*, and *Zinnia* seeds at 9.2 kV discharge voltage and 0.2 A discharge current and 1.49 W/cm^2 of discharge power density. The growth enhancement was 250, 80, 60, 30, and 20% for *Raphanus sativus*, *Oryza sativa*, *Arabidopsis Thaliana*, *Plumeria*, and *Zinnia*, respectively, after scalar DBD treatment [39,97]. Similar scalar DBD used to treat *Arabidopsis thaliana* seeds see Figure 5a, which results in accelerated growth, shorter harvest time, increased total seed weight, and increased seed number, as shown in Figure 5b [46]. Further, Kitazaki et al. also used the same scalar DBD plasma to analyze the growth of radish sprouts (*Raphanus sativus* L.) using combinatorial analysis. Authors observed 250% growth enhancement when seeds were placed at x = 5 and y = 3 mm [40]. In a separate study, this scalar DBD treated the radish sprouts (*Raphanus sativus*) seeds for 3 min in the presence of different feed gases like He, Ar, N_2, Air, O_2, and NO (10%) + N_2. For He, N_2, and Ar feeding gases, plasma treatment showed a limited influence on plant growth; however, for O_2, Air and NO (10%) + N_2 gases plasma had significant on growth enhancement [41].

Hayashi et al. treated the seeds of radish sprouts (*Raphanus sativus* var. *longipinnatus*) with plasma torch for 60 min. Plasma device with O_2 and air feeding gas had a frequency of 12 kHz with a varied applied voltage of 7–10 kV. The total length (stem and root length) increased by 1.6 times for O_2-plasma and 1.2 times for air-plasma treatment than control (without plasma treatment) [98]. In another study, the radish sprouts seeds (*Raphanus sativus*) treated with an Ar-plasma jet. Plasma treatment at 140 W power results in increased total mass by 9–12% and increased average length by 3 cm in comparison with untreated seeds [42]. In a separate study, surface discharge plasma treatment of radish sprouts (*Raphanus sativus*) seeds for 20 min results in no change in the germination dynamics. The surface discharge plasma operated in AC mode with the sinusoidal voltage of 15 kV at 50 Hz frequency with an average power of 2.7 W. However, root and sprout lengths were increased by 11 and 10%, and root and sprout weight were increased to 30 and 15%, respectively, after plasma treatment [43].

Substantial the effect of plasma treatment on wheat seeds studied by both low-pressure plasma and atmospheric pressure plasma. Dobrin et al. treated the wheat seeds (*Triticum aestivum* L.) with surface discharge plasma with airflow of 1 L/min had a sinusoidal voltage of 15 kV at 50 Hz frequency with an average power of 2.7 W for 15 min. The plasma treatment had a small effect on the germination rate, but at the same time, plasma treatment showed a pivotal impact on growth parameters. The root-to-shoot (R/S) ratio was higher for treated samples than untreated samples [51]. Meng et al. treated the wheat seeds (Xiaoyan 22) with DBD plasma had discharge voltage 0–50 kV and frequency of 50 Hz with different working gases like Air, N_2, Ar, and O_2. After plasma treatment, germination potential increased to 24, 28, and 35.5% for Air, N_2, and Ar feeding gases plasma, respectively, as compared to control [52].

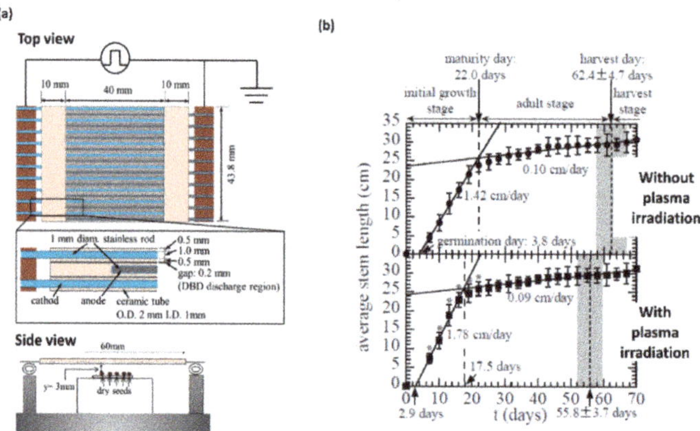

Figure 5. (**a**) Schematic depiction of scalable dielectric barrier discharge plasma device (**b**) Average stem length for plants as a function of time with and without plasma treatment. Reproduced from reference [46], Copyright (2016) The Japan Society of Applied Physics.

Li, et al. used the Air-DBD (1.50 W discharge power and 13.0 kV discharge voltage) to treat the wheat seed (Xiaoyan 22) for 7 min. The authors observed that the germination rate, germination potential, germination index, and vigor index were increased by 9.1, 26.7, 16.9, and 46.9%, respectively [53]. Further, the same group studied the adverse effects of drought stress on wheat seed (Xiaoyan 22) germination and seedling growth in the presence of the DBD as mentioned earlier. The germination potential and germination rate were increased to 27.2 and 27.6%, respectively, after plasma treatment. Additionally, root and shoot length increased after DBD treatment [54]. Recently, Lotfy et al. used the N_2-plasma jet at discharge voltage and discharge current of 2.6 kV and 38.1 mA, respectively, to treat wheat seeds (Giza 168). The N_2-plasma jet treatment for 4 min results in 54.3% higher mean dry weight than control samples [55].

Sunflower seeds and sprouts are in high demand due to its application in food industries, biofuels, cosmetics, and lubricants. Therefore, researchers are using plasma to enhance the productivity of sunflowers. Sunflower (*Helianthus annuus*) seeds treated with scalar DBD for 7 and 11 min at a discharge voltage of 9.2 kV, and the current of 0.2 A, due to plasma treatment the adverse effect on germination kinetics was observed. Although root length was increased by 44%, and the weight of seedlings and leaves were increased by 16 and 15%, respectively. Moreover, DBD plasma treatment for 11 min did not affect root length and sprout weight. However, DBD treatment results increased leaf weight and leaf number by 15 and 10%, respectively, with respect to control. In addition, no effect on the stem length of sunflower seedlings was observed [56]. Matra et al. treated the same sunflower *(Helianthus annuus* L.) seeds with NTP had average discharge voltage and current of 7.451 kV and 0.073 mA, respectively, with Ar: O_2 feeding gases. After plasma treatment, the dry weight was 1.79 times heavier, and the average shoot length was 2.69 times higher, than control [57]. A recent study showed the DBD treatment effect on sunflower (*Helianthus annuus*) seeds for 120 s at 90 W power, which results in the enhancement in sunflower seed germination [58].

Diffuse coplanar surface barrier discharge (DCSBD) used to treat the Pea (*Pisum sativum*) seeds with 14 kHz frequency, 10 kV sinusoidal voltage, 370 W input energy, and 2.3 W cm^{-2} of average power density. After the DCSBD plasma treatment for 120 s, the total percentage of seed germination significantly increased to 95%, whereas the germination percentage was 77.5% for control. Similarly, the length of the shoot and root, as well as dry seedling weight also increased as compared to control samples after DCSBD treatment [61]. Khatami and Ahmadinia, treated pea seeds for 30 and 60 s with a plasma device (self-made FSG plasma (a semi-automatic device)) system with an applied voltage of 15 kV and air gas flow of 5 L/min. After 30 and 60-s plasma treatment, 80 and 74% of seeds, respectively, were germinated, although, in the control sample, only 40% seeds were germinated, after 14 days [62].

Zhou et al. used a micro-plasma array with an AC frequency of 9.0 kHz to treat Mung bean seeds in aqueous media in the presence of different feed gases such as Air, O_2, N_2, and He. The authors observed that seed germination and growth of mung bean were dependent on the feed gases and treatment time. Air micro-plasma array treatment improved the seed germination rate and seedling growth in comparison to other feeding gases plasma (O_2, N_2, and He) treatment. Air and O_2 micro-plasma array treatments substantially improved the germination index by 58.3% and 41.7%, respectively, whereas no significant difference observed for He or N_2 plasma as compared to control [63]. Pérez-Pizá et al. treated the soybean seeds with DBD plasma with AC power supply (0–25 kV) operating at 50 Hz with N_2 or O_2 as carrying gases. N_2 and O_2 plasma treatment for 3- and 2-min results in 1.2-fold incremented in total fresh weight than control, and the full length of soybean plant increased to 4–10% after plasma treatment [60].

Zahoranová et al. used DCSBD plasma (operated at 14 kHz frequency, 20 kV (peak-to-peak) sinusoidal high voltage, and 80 W cm^{-3} power density with an input power of 400 W) to treat maize (*Zea mays* L.; cv. Ronaldinio) seeds. After 60- and 120-s plasma treatment, no significant difference in germination was noted in comparison to untreated maize seeds. However, for 300 s of DCSBD treatment, the germination rate was decreased to 7%. However, the vigor index increased by 23% for 60-s plasma treatment compared to control [99]. Further, rice (*Oryza sativa)* seeds were treated with hybrid cold plasma (HCP) (high voltage of 14 kV_{pp}, 700 Hz frequency, and 4.8 W power (Matsushita Electronic components)), working in Ar gas. The final germination was 98% for plasma-treated rice seeds, whereas 90% for control. However, seedling length remains the same for both non-treated and treated rice seeds [100]. Amnuaysin et al. treated the rice (*Oryza sativa* L.) seeds with Air-DBD operated at 5.5 kHz frequency for 60 s. The germination rate was 92.67% after plasma treatment, while 85% for control. Additionally, the vigor index, germination rate, seedling growth, fresh weight (root and shoot), and dry weight (root and shoot) showed improvement after plasma treatment [101].

Tomato seeds (*Lycopersicon esculentum*) hybrid Belle F1 treated with a coaxial DBD reactor with air as a working gas had a flow rate of 15 L/min with 50 Hz frequency and 1.43 W discharge power. They noted a 77% germination rate for 5 min DBD treatment while 68% for the control. Although 5 min DBD exposure results in three times increase of average root length than control seeds. Additionally, the root-to-shoot ratio (R/S) for control seeds was 0.51, whereas for 5- and 30-min plasma treatment, it was 0.87 and 0.73, respectively [73]. Pumpkin (*Cucurbita pepo*) treated with plasma jet with a high voltage pulsed DC system operated with pulse amplitude 8 kV, pulse frequency 6 kHz, pulse width 1 µs, and pulse rise and fall time ~70 ns. After the treatment, seed germination, and seed growth was enhanced [74]. Schnabel et al. showed that *Brassica napus* L. seeds treated with DBD had a frequency of 5.7 kHz and showed no significant change in seed germination than control. However, 10 min DBD treatment results in increased germination percentage by 3 and 13% after 24 and 48 h, respectively, compared with control seeds [75].

Andrographis paniculata seeds treated with DBD plasma at 4250 V for 10 s and 5950 V for 20-s results in enhanced seed germination [76]. Fenugreek (*Trigonella foenum-graecum*) seeds treated with Ar-plasma jet (applied voltage 16 kV and an applied frequency of 24 kHz). After plasma treatment, the seed germination rate improved by 7 and 4 times with and without an accelerating grounded electrode, respectively [77]. Mulungu (*Erythrina velutina*) seeds treated with He-DBD at 10 kV applied voltage, 750 Hz frequency, and 150 W power for different time intervals. At 60 s He-DBD treatment, the seed germination rate was 5% higher than control [78]. *Hybanthus calceolaria* seeds treated with He-plasma jet working at 8.1 kV discharge voltage, 720 Hz frequency for 1 min showed an increase in seed germination by 3.5 times than untreated seeds [79]. Molina et al. treated the Nasturtium seeds (*Tropaeolum majus* L.) with He-DBD plasma for 10- and 30-s results in increased germination to 68.3 and 61.7%, respectively, as compared with control seeds under drought conditions. However, more extended plasma treatment results in decreased germination efficiency [80]. Thuringian Mallow (*Lavatera thuringiaca*) seeds treated with GlidArc reactor with dry nitrogen as working gas at 50 Hz discharge frequency, 680 V applied root mean square (RMS) voltage, 33 mA RMS current and 40 W of mean power. The germination was 60% for both 2- and 5-min plasma treated seeds, while 36.25% for control seeds [81]. In another study, pre-treatment of hemp with Gliding arc working at 50 Hz power frequency at a flow rate of 10 L/min of humid air, resulted in increased length of seedlings, seedling accretion, and weight of seedling [69].

Cucumber (*Cucumis sativus*) and Pepper (*Capsicum annuum*) seeds treated with DCSBD plasma in ambient air, had 15 kHz frequency, and 400 W input power. After the DCSBD treatment, the germination percentage for both cucumber and pepper increased for short plasma treatment time, while decreased for more prolonged plasma exposure [82]. Mitra et al. treated the *Cicer arietinum* L. seeds with surface micro-discharge (SMD) plasma in ambient air had a power density of 10 mW/cm^2. After 1 min SMD treatment, the seed germination was 89.2% improved, and the mean germination time also decreased to 2.7 days than control [102]. In another study, high voltage nanosecond pulsed plasma (NPP) and micro DBD plasma used to treat the Spinach (*Spinacia oleracea*) seeds. NPP plasma had 6 kV and 0.7 kA of discharge voltage and current, respectively, with 0.3 J per one-shot pulse discharge energy. Whereas, micro DBD plasma had 6 kV, 14 mA, and 22 kHz of discharge voltage, current, and frequency, respectively, with Air and N_2 working gases. The authors observed that seeds treated with NPP showed 75–80% increased germination with one or five shots in comparison to 60% for untreated seeds. Whereas the germination decreases for ten shots of NPP. Seedlings growth and dry weight were increased after Air DBD treatment, while no substantial changes observed after N_2 DBD treatment [83].

Very recently, barley (*Hordeum vulgare*), chili pepper (*Capsicum*), black pine (*Pinus nigra*), and basil (*Ocimum basilicum*) seeds treated with different types of plasma [44,84–86]. Song et al. treated the barley (*Hordeum vulgare* L.) seeds with surface dielectric barrier discharge (SDBD) plasma operated at 14.4 kHz driving frequency, 8 kV peak-to-peak voltage, and 51.7 W average power. The 6-min plasma exposure increased the fresh weight of whole barley seedlings by 137.5% [84]. Chili pepper seeds treated with Ar-plasma for 15 s with a high voltage DC pulse operated at different powers. Seed germination was 86.53, 100, 95.51, and 94.23% at 0.41, 0.48, 0.55, and 0.61 W power, respectively, whereas control was 24.35%. The seed germination percentage was highest at 0.48 W [44]. Black Pine (*Pinus nigra*) seeds treated with DCSBD plasma for 1–60 s with an input power of 400 W for each treatment period. The highest germination index was for 3 s and the smallest for 60 s of plasma treatment [85]. Basil (*Ocimum basilicum*) seeds treated with volume DBD in humid air, resulted in the rapid germination rate for 1 and 3 min DBD treatment as compared to control [86].

4. Effect of Plasma Treated Water on Seed Germination and Growth Enhancement

In the last few decades, plasma-treated water (PTW) showed the potential application in the agriculture and food industry [103–106]. The plasma exposer to water results in the production of different reactive oxygen and nitrogen species (RONS); therefore, PTW has a mixture of different RONS, mainly those that have a long lifetime, e.g., H_2O_2, NO_2^-, NO_3^-, etc. In this review, we will discuss the influence of PTW on the germination and growth of plants.

Gram (*Vigna mungo*) seeds treated with H_2O-O_2 discharge plasma generated PTW. H_2O-O_2 discharge plasma worked with high voltage (3–6 kV, 3–10 kHz) power supply. The PTW produced from H_2O-O_2 discharge plasma for 3, 6, 9, 12, and 15 min, revealed the significant improvement in seedling growth of black gram seeds was observed after PTW treatment as compared to control. PTW generated by 6 min plasma treatment showed the foremost cumulative germination, while 3 min plasma treatment produced PTW showed the higher vigor index. Seeds treated with PTW created after

3-, 6- and 9-min plasma treatment showed longer shoot and root as compared to untreated samples. Additionally, PTW produced after plasma treatment for 12 min displayed the highest dry weights of shoots among other PTW treated samples [87].

Sarinont et al. showed that PTW generated by scalable DBD device using various gases like Air, O_2, N_2, He, and Ar used to treat the Radish sprouts (*Raphanus sativus* L.) seeds with discharge frequency of 14 kHz. The power was 5.85, 4.51, 3.54, 8.95, and 9.64 W for air, O_2, N_2, He, and Ar working gases, respectively. The PTW kept for 1 h and 1 day after scalable DBD treatment at room temperature to minimize the effect of short-lived reactive species. The PTW kept for 1 h produced from Air, O_2, He, N_2, and Ar plasmas showed 1.62-, 1. 38-, 1.13-, 1.12-, and 1.04-times increased in seedling length of sprouts as compared to control. Although one day kept PTW produced from Air, O_2, He, N_2, and Ar plasmas showed 1.52, 1.28, 1.13, 1.10, and 1.08 times increased length of sprouts, respectively, more than control [88]. In another study, Sivachandiran and Khacef treated the Radish sprouts (*Raphanus sativus* L.) seeds with PTW created from double dielectric barrier discharge reactors (DBD) for 15- and 30-min treatment. The PTW generated from the double DBD discharge plasma (high voltage pulsed power supply (40 kV, 1 kHz)) showed enhance seeds germination rate and seedling growth. The PTW produced from 15- to 30-min plasma treatment had 60 and 100% germination rate as compared to control (40%). They also noticed that PTW produce from 15- to 30-min plasma treatment after three days had an average seedling length of 5 mm and 15 mm, respectively [89].

Soybeans (*Glycine max*) seeds were treated with PTW obtained from NTP in Air at voltage 80 kV and a frequency of 50 Hz in AC. The soybean seed germination was enhanced by PTW treatment in comparison with untreated seeds [90]. Very recently, Mung bean seeds treated with PTW produced at atmospheric pressure in specially designed two electrochemical cells using HV-DC power supply. The plasma discharged with circulating water for 3 h. No significant difference was observed on the growth rate of mung bean sprouts for control and cathodic electrochemical cell water. However, the growth rate was significantly lower for anodic electrochemical cell water as compared to control [91].

Zinnia annual (*Zinnia elegance*) seeds treated with PTW produced from underwater electric front type discharged for 5 min, operated at a voltage not more than 1.5 kV, and discharge current was 50–70 mA. Germinability rate increased by 50%, and plant roots rose to 1.5- to 2-fold as compared to the control after PTW treatment [92]. *Brassica rapa* var. *perviridis* seeds treated with PTW generated by underwater discharge with a repetition rate of 250 pps had a peak voltage of 30 kV using pulsed power generator for 10- and 20-min. The authors noticed that the dry weight of the plant increased to 3.9 and 6.6 times, additionally, leaf length increased to 2.1 and 2.5 times after PTW treatment generated by plasma for 10- and 20-min, respectively, as compared to control [93]. Zhang et al. treated the *Lentils* seeds with PTW created with He-APPJ, showed 80% germination rates, and for control was 30%. Additionally, they also noticed the higher stem elongation rates and final stem lengths in PTW treated samples than commercial fertilizer (control) [94].

Recently, Adhikari et al. treated the Tomato (*Solanum lycopersicum* L.) seeds with PTW generated from the plasma jet. The frequency, current, and applied voltage of the plasma jet discharge were 83.5 kHz, 70.39 mA, and 0.66 kV, respectively. PTW generated from plasma jet for 15- and 30-min treatment showed better morphological growth compared to control seedlings. More PTW produced at 15- and 30-min plasma jet treatment showed higher shoot and root length, but there is no significant change observed for the PTW generated after 60 min plasma treatment [95]. In another recent study on Rapeseed (*Brassica napus*) seeds showed a significant improvement in germination rate and seedling vigor as compared with control when treated with PTW produced with Ar and O_2 feed gases [96].

5. Patents Related to Seed Germination and Seed Growth Using Low-Pressure/Medium-Pressure/Atmospheric-Pressure Plasma

A few dozen patents using discharge plasma for plant seed to induce seed-activity have been taken out according to PATENTSCOPE by WIPO [107]. Typically, one of those is direct irradiation. Masaru et al. unveil direct radiation of the atmospheric-pressure plasma jet operated at 50,662–202,650 Pa pressure on rice (*Oryza sativa* L.) seeds, which results in growth improvement [108].

The working gas was Ar or inert gas; the plasma density was 1×10^{14}–1×10^{17} /cm^3, and the plasma temperature was 1000–2500 K. According to the Masaru group, the preferable value of multiplication of plasma density with irradiation time was 6×10^{16}–2×10^{17} s/cm^3 for a positive effect. Additionally, Hayashi et al. showed the improved method for the growth of radish (*Raphanus Sativus* L.) by plasma irradiation to dry seeds [109]. The DBD type plasma generated in stainless chamber φ200 mm × 450 mm at 80 Pa. The feeding gas was O_2, the frequency and power of the plasma source were 13.56 MHz and 50 W, respectively, and the exposure time was 60 min. The whole stem and root length, and cotyledon width of seedling with the plasma irradiation were 167 mm, 70 mm, 97 mm, and 10.6 mm, and those of control was 120 mm, 41 mm, 79 mm, and 9.5 mm, respectively. They revealed that not only the stem and root length but also the amount of endogenous thiol increased with plasma-irradiation time by changing water vapor pressure from 30 to 270 Pa. Konstantinovich et al. combined the exposure of low-pressure plasma and activated water containing anolyte and catholyte. Low-pressure plasma worked at 0.01–0.1 W/cm^3 and ≈26–150 Pa under inert gas, oxygen, nitrogen, or a mixture of oxygen and nitrogen environment at 20–40 °C for 10–45 s. The authors reduced the seed germination period of radish, pea, beet, cabbage, tomato, barley, lentils, pumpkin, corn, wheat, etc., by 2–4 days [110]. Another group used plasma at 184–188 W for 8–10 s, rested for 4–5 s, 103–107 W for 13–15 s, rested for 2–3 s, and 264–270 W for 11–13 s for black bean seeds; they observed the acceleration in the seed germination after plasma treatment [111]. Shi et al. used plasma seed at 51–53 W for 150–170 s, 141–145 W for 90–110 s, and 77–79 W for 190–210 s twice for dry pumpkin seeds, and at this treatment condition, the seed germination process was promoted [112]. Guo et al. used an atmospheric–pressure DBD plasma generated with a frequency of 3–4 kHz and a voltage of 4–7 kV to treat dry peony seed for 40–60 s [113]. Another research group irradiated the wheat seeds with low-pressure plasma with He as the working gas at a power of 1–500 W for 15–20 s. The highest germination rate of 12.72% observed at 80 W plasma power and irradiated for 18 s [114]. Chengdong et al. found that irradiation of an atmospheric-pressure DBD plasma generated at a frequency of 2.5–3.5 kHz and a voltage of 3.5–4.5 kV for 65–85 s to Chinese rosewood (*Dalbergia odorifera* T. Chen) seeds preceding soaked in the distillate of wood for several h, resulted in improved seed germination and subsequent growth [115]. Later, a group of researchers treated the peony seed by atmospheric–pressure DBD plasma with a frequency of 3–5 kHz and a voltage of 5–7 kV for 50–60 s, showed the improvement in germination [116]. On the other hand, Li et al. observed the increased seed viability and the crop yield by 6.5% after maturation compared to the control after plasma treatment working at 220 ± 22 V with a frequency of 50 Hz and a power of 55 W or less [117]. The reviewed patents emphasize the positive effects of plasma treatment on the improved seed germination.

6. Probable Mechanism and Future Perspectives

Overall research outputs by several research groups showed that the germination rate and growth of seedling had increased with direct plasma treatment (low/medium/atmospheric pressure plasma) or through PTW treatment. In this review, we discussed the probable mechanism of enhanced germination rate and growth of the seedling as well (as shown in Figure 6) as the possibility of genetic mutation in seeds observed by various research groups around the world. Hayashi et al. revealed active oxygen species produced from O_2-low pressure plasma-activated the genes in *Arabidopsis thaliana* seeds that were responsible for the cell elongation proteins. Moreover, in the second generation, these genes are not activated; this concluded that there was no genetic mutation in the seeds [45]. Further, the enhanced germination rate and growth of seedling mechanism were discussed below and in Figure 6.

Volin et al. showed that plasma treatment in different feeding gases could alter germination time [8]. Treatment of seeds in the presence of carbon tetrafluoride delays the germination of two pea cultivars (*Pisu sativum* cv. Little Marvel, *P. sativum* cv. Alaska) and radish (*Raphanus sativus*). At the same time, plasma treatment in the presence of cyclohexane can significantly accelerate the germination percentage of soybean [8]. Hence, the different feeding gas alters the plasma chemistry that provides alternatives for seed coating [8].

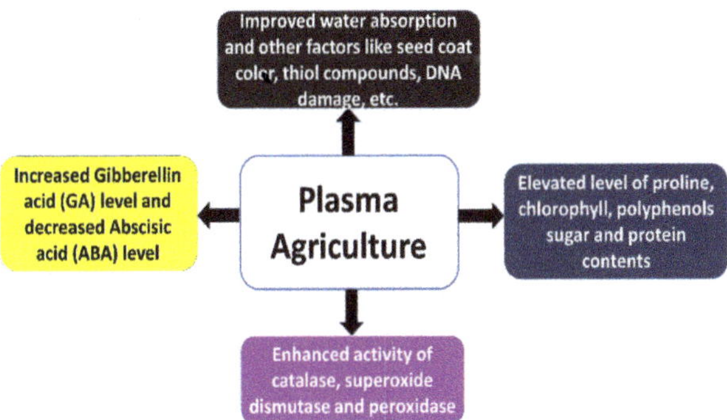

Figure 6. Possible mechanism in plasma agriculture.

Sarinont, et al. revealed an increase in chlorophyll and carotenoid concentration in Radish sprouts seeds after scalar DBD treatment [118]. Saberi et al. showed improvement in the photosynthesis rate, stomatal conductance, and chlorophyll content in low-pressure plasma-treated wheat seeds [48]. Ji et al. showed an increased level of chlorophyll and total polyphenols contents in spinach seedlings after air DBD treatment [83]. In a very recent report, Sajib et al. showed the increased chlorophyll content in black grams seeds after PTW treatment [87].

Phytohormones played an essential role in plant growth. Zukiene et al. showed increased/decreased gibberellin (GA)/abscisic acid (ABA) ratio that results in positive or negative effects on germination and/or seedling growth of sunflower seeds after treatment with scalar DBD [56]. Whereas Stolárik, et al. observed the changes in catabolites, conjugates and endogenous hormones (auxins and cytokinins) of pea seedlings after treatment with DCSBD plasma [61]. In another study, Degutyte-Fomins et al. showed a decrease and increase in ABA and GA contents in radish sprouts, respectively, after DBD treatment [119]. Ji et al. observed an elevated level of GA3 hormone in spinach seedlings after NPP treatment [83].

On the other hand, Li et al. showed an elevated level of proline, soluble sugar contents, and osmotic-adjustment products after DBD treatment on wheat seeds [53]. Another study showed that low-pressure plasma treatment on the soybean seeds results in increased protein and soluble sugar content [59]. Whereas, Ling et al. observed the elevated level of soluble sugar and protein contents, whereas reduction in malondialdehyde content after low-pressure plasma treatment on oilseed rape seeds [68]. Similarly, Guo et al. observed improvement in proline and soluble sugar contents and decreased the malondialdehyde content in wheat seeds treated with DBD [54].

Hosseini et al. showed improved seed's water uptake in artichoke seeds after low-pressure N_2 plasma treatment [65]. Meng et al. showed that DBD treatment on wheat seeds results in an improvement in the capacity for water absorption and activation of several physiological reactions [52]. Bormashenko et al. showed that low-pressure plasma treatment noticeably increased the water imbibition in bean seeds through the testa, it is independent of the micropyle effect [64]. Alves Junior et al. observed the change in hydrophilicity and water absorption of Mulungu seeds after plasma jet treatment [78]. Volkov et al. showed that pumpkin seeds treated with plasma jet showed structural deformations of seeds such as surface defects, and hydrophilic pores that results in enhanced water uptake [74]. Fadhlalmawla et al. reported that Fenugreek seeds treatment with cold atmospheric pressure plasma jet showed increased water imbibition and absorption that was due to etching on seed coat surfaces [77]. Additionally, Li et al. showed that improvement in the water absorption capacity of wheat seed after DBD treatment is due to the etching effect [53]. Saberi et al. demonstrated the improved tolerance of wheat plants against the haze and increased relative water content after low-pressure treatment [48].

Further, it was showed that the production of reduction-type thiol compound changed in radish sprout seeds after treated with plasma torch; these thiol compounds were responsible for growth regulation mechanisms in plants [98]. Hayashi et al. mention that thiol content in seeds with plasma irradiation may be associated with plant growth. Moreover, the redox properties of plasma irradiation as a function of water vapor pressure results in the variation of cysteine peak values at 2587 cm^{-1} (thiol group; –S–S–) and 520 cm^{-1} (disulfide bond; –SH) using Fourier transform infrared spectroscopy (FTIR). The increased amount of thiol is due to the reduction of cystine by active hydrogen. The increased cysteic acid was due to oxidative modification of cysteine during plasma irradiation [109]. Very recently, Song, et al. demonstrated the enhanced contents of the primary metabolites, especially the free amino acids and soluble sugars, as well as secondary metabolites like phytochemicals, e.g., saponarin, GABA, and policosanols in barley seeds after treated with SDBD plasma [84].

Other observations in plasma treated seeds by different authors such as Kyzek et al. noticed the DNA damage in Pea seeds due to DCSBD plasma treatment [120]. Wang et al. observed the etching in cotton seeds after DBD treatment [121]. Whereas Adhikari et al. found the elevated expression of pathogenesis-related (PR) gene and defense hormones in tomato seeds treated with PTW [95]. Hosseini et al. and Tong et al. observed the increase in catalase activity after DBD and low-pressure treatment [65,76]. Additionally, Li et al. showed increased activities of superoxide dismutase and peroxidase [53]. In a very recent study, Koga et al. noticed that different seed coat color of radish sprouts responds differently to the scalar DBD treatment. Hence, pigments in the seed coat play a vital role in plasma treatment [122].

All the above studies showed that plasma treatment and PTW treatment had a positive effect on seeds; it improves the germination percentage, seedling growth, and yield. Although these results are laboratory-based; hence, the real field study will be more helpful, which can open new prospects in farming. Ahn et al. conducted the actual field experiments [123]. Authors treated the yellow dent corn hybrid seeds with different types of plasma-like low-pressure RF plasma, microwave-driven plasma (atmospheric pressure), and DBD plasma (atmospheric pressure). Corn seeds treated by low-pressure RF (13.56 MHz) plasma at ≈13.3 Pa with N_2 as process gas. Separately, corn seeds treated with a microwave atmospheric plasma jet with 500 W microwave power. DBD plasma with He as feed gas treated the corn seeds with a high voltage of 15 kV at 35 kHz. The yield obtained for low-pressure RF was relatively higher than other plasma systems but not as high as the control. The authors concluded that there was no statistically significant difference in the yield found between the control and plasma treatment [123].

7. Conclusions

In this review, we concluded that the direct plasma treatments working at low/medium/atmospheric pressure and PTW treatments could change the physical and biochemical properties of seeds. These changes result in the enhancement of seed germination and seedling growth. However, it was imperative to do the real field experiments with the plasma-treated seeds to make it useful to society. Otherwise, plasma agriculture will be limited to the research articles and will not be of value to society.

Author Contributions: P.A. designed and wrote the review. K.I. and M.S. checked the contents, K.K. provided the figures and T.O. compile the patent data for the study. All authors have read and agreed to the published version of the manuscript.

Funding: This work is supported by JSPS-KAKENHI grant number 20K14454. Additionally, partly supported by JSPS KAKENHI Grant Number JP16H03895, JP19H05462, JP19K14700, JP20H01893, JP20K14454, Plasma Bio Consortium, and Center for Low-temperature Plasma Sciences, Nagoya University.

Conflicts of Interest: The authors declare no conflict of interest.

References

1. Heydecker, W.; Coolbear, P. Seed treatments for improved performance survey and attempted prognosis. *Seed Sci. Technol.* **1977**, *5*, 353–426.
2. Taylor, A.G.; Allen, P.S.; Bennett, M.A.; Bradford, K.J.; Burris, J.S.; Misra, M.K. Seed enhancements. *Seed Sci. Res.* **1998**, *8*, 245–256. [CrossRef]
3. Halmer, P. Seed Technology and Seed Enhancement. *Acta Hortic.* **2008**, *771*, 17–26. [CrossRef]

4. Elsayed, B.B.; Hassan, M.M.; El Ramady, H.R. Phylogenetic and characterization of salt-tolerant rhizobial strain nodulating faba bean plants. *Afr. J. Biotechnol.* **2013**, *12*, 4324–4337. [CrossRef]
5. Araújo, S.D.; Paparella, S.; Dondi, D.; Bentivoglio, A.; Carbonera, D.; Balestrazzi, A. Physical methods for seed invigoration: Advantages and challenges in seed technology. *Front. Plant Sci.* **2016**, *7*, 646. [CrossRef]
6. Šerá, B.; Špatenka, P.; Šerý, M.; Vrchotová, N.; Hrušková, I. Influence of plasma treatment on wheat and oat germination and early growth. *IEEE Trans. Plasma Sci.* **2010**, *38*, 2963–2968. [CrossRef]
7. Dubinov, A.E.; Lazarenko, E.M.; Selemir, V.D. Effect of glow discharge air plasma on grain crops seed. *IEEE Trans. Plasma Sci.* **2000**, *28*, 180–183. [CrossRef]
8. Volin, J.C.; Denes, F.S.; Young, R.A.; Park, S.M.T. Modification of seed germination performance through cold plasma chemistry technology. *Crop Sci.* **2000**, *40*, 1706–1718. [CrossRef]
9. Straňák, V.; Tichý, M.; Kříha, V.; Scholtz, V.; Šerá, B.; Špatenka, P. Technological applications of surfatron produced discharge. *J. Optoelectron. Adv. Mater.* **2007**, *9*, 852–857.
10. Attri, P.; Razzokov, J.; Yusupov, M.; Koga, K.; Shiratani, M.; Bogaerts, A. Influence of osmolytes and ionic liquids on the Bacteriorhodopsin structure in the absence and presence of oxidative stress: A combined experimental and computational study. *Int. J. Biol. Macromol.* **2020**, *148*, 657–665. [CrossRef]
11. Attri, P.; Kim, M.; Choi, E.H.; Cho, A.E.; Koga, K.; Shiratani, M. Impact of an ionic liquid on protein thermodynamics in the presence of cold atmospheric plasma and gamma rays. *Phys. Chem. Chem. Phys.* **2017**, *19*, 25277–25288. [CrossRef] [PubMed]
12. Attri, P.; Kim, M.; Sarinont, T.; Ha Choi, E.; Seo, H.; Cho, A.E.; Koga, K.; Shiratani, M. The protective action of osmolytes on the deleterious effects of gamma rays and atmospheric pressure plasma on protein conformational changes. *Sci. Rep.* **2017**, *7*, 8698. [CrossRef] [PubMed]
13. Hoon Park, J.; Kumar, N.; Hoon Park, D.; Yusupov, M.; Neyts, E.C.; Verlackt, C.C.W.; Bogaerts, A.; Ho Kang, M.; Sup Uhm, H.; Ha Choi, E.; et al. A comparative study for the inactivation of multidrug resistance bacteria using dielectric barrier discharge and nano-second pulsed plasma. *Sci. Rep.* **2015**, *5*, 13849. [CrossRef] [PubMed]
14. Attri, P.; Kim, Y.H.; Park, D.H.; Park, J.H.; Hong, Y.J.; Uhm, H.S.; Kim, K.-N.; Fridman, A.; Choi, E.H. Generation mechanism of hydroxyl radical species and its lifetime prediction during the plasma-initiated ultraviolet (UV) photolysis. *Sci. Rep.* **2015**, *5*, 9332. [CrossRef]
15. Attri, P.; Han, J.; Choi, S.; Choi, E.H.; Bogaerts, A.; Lee, W. CAP modifies the structure of a model protein from thermophilic bacteria: Mechanisms of CAP-mediated inactivation. *Sci. Rep.* **2018**, *8*, 10218. [CrossRef]
16. Choi, S.; Attri, P.; Lee, I.; Oh, J.; Yun, J.-H.; Park, J.H.; Choi, E.H.; Lee, W. Structural and functional analysis of lysozyme after treatment with dielectric barrier discharge plasma and atmospheric pressure plasma jet. *Sci. Rep.* **2017**, *7*, 1027. [CrossRef]
17. Attri, P.; Bogaerts, A. Perspectives of Plasma-treated Solutions as Anticancer Drugs. *Anticancer Agents Med. Chem.* **2019**, *19*, 436–438. [CrossRef]
18. Attri, P. Cold Atmospheric Plasma Activated Solution: A New Approach for Cancer Treatment. *Anticancer Agents Med. Chem.* **2018**, *18*, 768. [CrossRef]
19. Attri, P.; Choi, E.H. Influence of Reactive Oxygen Species on the Enzyme Stability and Activity in the Presence of Ionic Liquids. *PLoS ONE* **2013**, *8*, e75096. [CrossRef]
20. Attri, P.; Sarinont, T.; Kim, M.; Amano, T.; Koga, K.; Cho, A.E.; Choi, E.H.; Shiratani, M. Influence of ionic liquid and ionic salt on protein against the reactive species generated using dielectric barrier discharge plasma. *Sci. Rep.* **2015**, *5*, 17781. [CrossRef]
21. Attri, P.; Park, J.H.; Ali, A.; Choi, E.H. How Does Plasma Activated Media Treatment Differ From Direct Cold Plasma Treatment? *Anticancer Agents Med. Chem.* **2018**, *18*, 805–814. [CrossRef] [PubMed]
22. Attri, P.; Tochikubo, F.; Park, J.H.; Choi, E.H.; Koga, K.; Shiratani, M. Impact of Gamma rays and DBD plasma treatments on wastewater treatment. *Sci. Rep.* **2018**, *8*, 2926. [CrossRef] [PubMed]
23. Shaw, P.; Kumar, N.; Kwak, H.S.; Park, J.H.; Uhm, H.S.; Bogaerts, A.; Choi, E.H.; Attri, P. Bacterial inactivation by plasma treated water enhanced by reactive nitrogen species. *Sci. Rep.* **2018**, *8*, 11268. [CrossRef] [PubMed]
24. Park, J.H.; Kim, M.; Shiratani, M.; Cho, A.E.; Choi, E.H.; Attri, P. Variation in structure of proteins by adjusting reactive oxygen and nitrogen species generated from dielectric barrier discharge jet. *Sci. Rep.* **2016**, *6*, 35883. [CrossRef] [PubMed]
25. Kumar, N.; Attri, P.; Choi, E.H.; Sup Uhm, H. Influence of water vapour with non-thermal plasma jet on the apoptosis of SK-BR-3 breast cancer cells. *RSC Adv.* **2015**, *5*, 14670–14677. [CrossRef]

26. Attri, P.; Yusupov, M.; Park, J.H.; Lingamdinne, L.P.; Koduru, J.R.; Shiratani, M.; Choi, E.H.; Bogaerts, A. Mechanism and comparison of needle-type non-thermal direct and indirect atmospheric pressure plasma jets on the degradation of dyes. *Sci. Rep.* **2016**, *6*, 34419. [CrossRef]
27. Sarangapani, C.; Patange, A.; Bourke, P.; Keener, K.; Cullen, P.J. Recent Advances in the Application of Cold Plasma Technology in Foods. *Annu. Rev. Food Sci. Technol.* **2018**, *9*, 609–629. [CrossRef] [PubMed]
28. Pankaj, S.K.; Bueno-Ferrer, C.; Misra, N.N.; Milosavljević, V.; O'Donnell, C.P.; Bourke, P.; Keener, K.M.; Cullen, P.J. Applications of cold plasma technology in food packaging. *Trends Food Sci. Technol.* **2014**, *35*, 5–17. [CrossRef]
29. Pankaj, S.K.; Keener, K.M. Cold plasma: Background, applications and current trends. *Curr. Opin. Food Sci.* **2017**, *16*, 49–52. [CrossRef]
30. Thirumdas, R.; Sarangapani, C.; Annapure, U.S. Cold Plasma: A novel Non-Thermal Technology for Food Processing. *Food Biophys.* **2015**, *10*, 1–11. [CrossRef]
31. López, M.; Calvo, T.; Prieto, M.; Múgica-Vidal, R.; Muro-Fraguas, I.; Alba-Elías, F.; Alvarez-Ordóñez, A. A review on non-thermal atmospheric plasma for food preservation: Mode of action, determinants of effectiveness, and applications. *Front. Microbiol.* **2019**, *10*, 622. [CrossRef]
32. Guo, Q.; Sun, D.-W.; Cheng, J.-H.; Han, Z. Microwave processing techniques and their recent applications in the food industry. *Trends Food Sci. Technol.* **2017**, *67*, 236–247. [CrossRef]
33. Randeniya, L.K.; de Groot, G.J.J.B. Non-Thermal Plasma Treatment of Agricultural Seeds for Stimulation of Germination, Removal of Surface Contamination and Other Benefits: A Review. *Plasma Process. Polym.* **2015**, *12*, 608–623. [CrossRef]
34. Ito, M.; Ohta, T.; Hori, M. Plasma agriculture. *J. Korean Phys. Soc.* **2012**, *60*, 937–943. [CrossRef]
35. Ito, M.; Oh, J.-S.; Ohta, T.; Shiratani, M.; Hori, M. Current status and future prospects of agricultural applications using atmospheric-pressure plasma technologies. *Plasma Process. Polym.* **2018**, *15*, 1700073. [CrossRef]
36. Puač, N.; Gherardi, M.; Shiratani, M. Plasma agriculture: A rapidly emerging field. *Plasma Process. Polym.* **2018**, *15*, 1700174. [CrossRef]
37. Kitazaki, S.; Koga, K.; Shiratani, M.; Hayashi, N. Growth Enhancement of Radish Sprouts Induced by Low Pressure O_2 Radio Frequency Discharge Plasma Irradiation. *Jpn. J. Appl. Phys.* **2012**, *51*, 01AE01. [CrossRef]
38. Hayashi, N.; Ono, R.; Uchida, S. Growth Enhancement of Plant by Plasma and UV Light Irradiation to Seeds. *J. Photopolym. Sci. Technol.* **2015**, *28*, 445–448. [CrossRef]
39. Sarinont, T.; Amano, T.; Kitazaki, S.; Koga, K.; Uchida, G.; Shiratani, M.; Hayashi, N. Growth enhancement effects of radish sprouts: Atmospheric pressure plasma irradiation vs. heat shock. *J. Phys. Conf. Ser.* **2014**, *518*, 012017. [CrossRef]
40. Kitazaki, S.; Sarinont, T.; Koga, K.; Hayashi, N.; Shiratani, M. Plasma induced long-term growth enhancement of *Raphanus sativus* L. using combinatorial atmospheric air dielectric barrier discharge plasmas. *Curr. Appl. Phys.* **2014**, *14*, S149–S153. [CrossRef]
41. Sarinont, T.; Amano, T.; Attri, P.; Koga, K.; Hayashi, N.; Shiratani, M. Effects of plasma irradiation using various feeding gases on growth of *Raphanus sativus* L. *Arch. Biochem. Biophys.* **2016**, *605*, 129–140. [CrossRef] [PubMed]
42. Matra, K. Non-thermal Plasma for Germination Enhancement of Radish Seeds. *Procedia Comput. Sci.* **2016**, *86*, 132–135. [CrossRef]
43. Mihai, A.L.; Dobrin, D.; Popa, M.E.; Mihai, A.L.; Dobrin, D.; Măgureanu, M.; Popa, M.E. Positive effect of non-thermal plasma treatment on radish. *Rom. Rep. Phys.* **2014**, *66*, 1110–1117.
44. Thisawech, M.; Saritnum, O.; Sarapirom, S.; Prakrajang, K.; Phakham, W. Effects of plasma technique and gamma irradiation on seed germination and seedling growth of chili pepper. *Chiang Mai J. Sci.* **2020**, *47*, 73–82.
45. Hayashi, N.; Ono, R.; Nakano, R.; Shiratani, M.; Tashiro, K.; Kuhara, S.; Yasuda, K.; Hagiwara, H. DNA microarray analysis of plant seeds irradiated by active oxygen species in oxygen plasma. *Plasma Med.* **2016**, *6*, 459–471. [CrossRef]
46. Koga, K.; Thapanut, S.; Amano, T.; Seo, H.; Itagaki, N.; Hayashi, N.; Shiratani, M. Simple method of improving harvest by nonthermal air plasma irradiation of seeds of *Arabidopsis thaliana* (L.). *Appl. Phys. Express* **2016**, *9*, 016201. [CrossRef]
47. Roy, N.C.; Hasan, M.M.; Talukder, M.R.; Hossain, M.D.; Chowdhury, A.N. Prospective Applications of Low Frequency Glow Discharge Plasmas on Enhanced Germination, Growth and Yield of Wheat. *Plasma Chem. Plasma Process.* **2018**, *38*, 13–28. [CrossRef]

48. Saberi, M.; Modarres-Sanavy, S.A.M.; Zare, R.; Ghomi, H. Improvement of photosynthesis and photosynthetic productivity of winter wheat by cold plasma treatment under haze condition. *J. Agric. Sci. Technol.* **2020**, *21*, 1889–1904.
49. Iqbal, T.; Farooq, M.; Afsheen, S.; Abrar, M.; Yousaf, M.; Ijaz, M. Cold plasma treatment and laser irradiation of *Triticum* spp. seeds for sterilization and germination. *J. Laser Appl.* **2019**, *31*, 042013. [CrossRef]
50. Jiang, J.; He, X.; Li, L.; Li, J.; Shao, H.; Xu, Q.; Ye, R.; Dong, Y. Effect of Cold Plasma Treatment on Seed Germination and Growth of Wheat. *Plasma Sci. Technol.* **2014**, *16*, 54–58. [CrossRef]
51. Dobrin, D.; Magureanu, M.; Mandache, N.B.; Ionita, M.-D. The effect of non-thermal plasma treatment on wheat germination and early growth. *Innov. Food Sci. Emerg. Technol.* **2015**, *29*, 255–260. [CrossRef]
52. Meng, Y.; Qu, G.; Wang, T.; Sun, Q.; Liang, D.; Hu, S. Enhancement of Germination and Seedling Growth of Wheat Seed Using Dielectric Barrier Discharge Plasma with Various Gas Sources. *Plasma Chem. Plasma Process.* **2017**, *37*, 1105–1119. [CrossRef]
53. Li, Y.; Wang, T.; Meng, Y.; Qu, G.; Sun, Q.; Liang, D.; Hu, S. Air Atmospheric Dielectric Barrier Discharge Plasma Induced Germination and Growth Enhancement of Wheat Seed. *Plasma Chem. Plasma Process.* **2017**, *37*, 1621–1634. [CrossRef]
54. Guo, Q.; Wang, Y.; Zhang, H.; Qu, G.; Wang, T.; Sun, Q.; Liang, D. Alleviation of adverse effects of drought stress on wheat seed germination using atmospheric dielectric barrier discharge plasma treatment. *Sci. Rep.* **2017**, *7*, 16680. [CrossRef]
55. Lotfy, K.; Al-Harbi, N.A.; Abd El-Raheem, H. Cold Atmospheric Pressure Nitrogen Plasma Jet for Enhancement Germination of Wheat Seeds. *Plasma Chem. Plasma Process.* **2019**, *39*, 897–912. [CrossRef]
56. Zukiene, R.; Nauciene, Z.; Januskaitiene, I.; Pauzaite, G.; Mildaziene, V.; Koga, K.; Shiratani, M. Dielectric barrier discharge plasma treatment-induced changes in sunflower seed germination, phytohormone balance, and seedling growth. *Appl. Phys. Express* **2019**, *12*, 126003. [CrossRef]
57. Matra, K. Atmospheric non-thermal argon–oxygen plasma for sunflower seedling growth improvement. *Jpn. J. Appl. Phys.* **2018**, *57*, 01AG03. [CrossRef]
58. Yawirach, S.; Sarapirom, S.; Janpong, K. The effects of dielectric barrier discharge atmospheric air plasma treatment to germination and enhancement growth of sunflower seeds. *J. Phys. Conf. Ser.* **2019**, *1380*, 12148. [CrossRef]
59. Li, L.; Jiang, J.; Li, J.; Shen, M.; He, X.; Shao, H.; Dong, Y. Effects of cold plasma treatment on seed germination and seedling growth of soybean. *Sci. Rep.* **2014**, *4*, 5859. [CrossRef]
60. Pérez-Pizá, M.C.; Cejas, E.; Zilli, C.; Prevosto, L.; Mancinelli, B.; Santa-Cruz, D.; Yannarelli, G.; Balestrasse, K. Enhancement of soybean nodulation by seed treatment with non–thermal plasmas. *Sci. Rep.* **2020**, *10*, 4917. [CrossRef]
61. Stolárik, T.; Henselová, M.; Martinka, M.; Novák, O.; Zahoranová, A.; Černák, M. Effect of Low-Temperature Plasma on the Structure of Seeds, Growth and Metabolism of Endogenous Phytohormones in Pea (*Pisum sativum* L.). *Plasma Chem. Plasma Process.* **2015**, *35*, 659–676. [CrossRef]
62. Khatami, S.; Ahmadinia, A. Increased germination and growth rates of pea and Zucchini seed by FSG plasma. *J. Theor. Appl. Phys.* **2018**, *12*, 33–38. [CrossRef]
63. Zhou, R.; Zhou, R.; Zhang, X.; Zhuang, J.; Yang, S.; Bazaka, K.; Ostrikov, K.K. Effects of Atmospheric-Pressure N_2, He, Air, and O_2 Microplasmas on Mung Bean Seed Germination and Seedling Growth. *Sci. Rep.* **2016**, *6*, 32603. [CrossRef] [PubMed]
64. Bormashenko, E.; Shapira, Y.; Grynyov, R.; Whyman, G.; Bormashenko, Y.; Drori, E. Interaction of cold radiofrequency plasma with seeds of beans (*Phaseolus vulgaris*). *J. Exp. Bot.* **2015**, *66*, 4013–4021. [CrossRef] [PubMed]
65. Hosseini, S.I.; Mohsenimehr, S.; Hadian, J.; Ghorbanpour, M.; Shokri, B. Physico-chemical induced modification of seed germination and early development in artichoke (*Cynara scolymus* L.) using low energy plasma technology. *Phys. Plasmas* **2018**, *25*, 013525. [CrossRef]
66. Gholami, A.; Safa, N.N.; Khoram, M.; Hadian, J.; Ghomi, H. Effect of Low-Pressure Radio Frequency Plasma on Ajwain Seed Germination. *Plasma Med.* **2016**, *6*, 389–396. [CrossRef]
67. Šerá, B.; Gajdová, I.; Šerý, M.; Špatenka, P. New Physicochemical Treatment Method of Poppy Seeds for Agriculture and Food Industries. *Plasma Sci. Technol.* **2013**, *15*, 935–938. [CrossRef]
68. Ling, L.; Jiangang, L.; Minchong, S.; Chunlei, Z.; Yuanhua, D. Cold plasma treatment enhances oilseed rape seed germination under drought stress. *Sci. Rep.* **2015**, *5*, 13033. [CrossRef]

69. Sera, B.; Sery, M.; Gavril, B.; Gajdova, I. Seed Germination and Early Growth Responses to Seed Pre-treatment by Non-thermal Plasma in Hemp Cultivars (*Cannabis sativa* L.). *Plasma Chem. Plasma Process.* **2017**, *37*, 207–221. [CrossRef]
70. Holc, M.; Primc, G.; Iskra, J.; Titan, P.; Kovač, J.; Mozetič, M.; Junkar, I. Effect of Oxygen Plasma on Sprout and Root Growth, Surface Morphology and Yield of Garlic. *Plants* **2019**, *8*, 462. [CrossRef]
71. Singh, R.; Prasad, P.; Mohan, R.; Verma, M.K.; Kumar, B. Radiofrequency cold plasma treatment enhances seed germination and seedling growth in variety CIM-Saumya of sweet basil (*Ocimum basilicum* L.). *J. Appl. Res. Med. Aromat. Plants* **2019**, *12*, 78–81. [CrossRef]
72. Billah, M.; Sajib, S.A.; Roy, N.C.; Rashid, M.M.; Reza, M.A.; Hasan, M.M.; Talukder, M.R. Effects of DBD air plasma treatment on the enhancement of black gram (*Vigna mungo* L.) seed germination and growth. *Arch. Biochem. Biophys.* **2020**, *681*, 108253. [CrossRef] [PubMed]
73. Măgureanu, M.; Sîrbu, R.; Dobrin, D.; Gîdea, M. Stimulation of the Germination and Early Growth of Tomato Seeds by Non-thermal Plasma. *Plasma Chem. Plasma Process.* **2018**, *38*, 989–1001. [CrossRef]
74. Volkov, A.G.; Hairston, J.S.; Patel, D.; Gott, R.P.; Xu, K.G. Cold plasma poration and corrugation of pumpkin seed coats. *Bioelectrochemistry* **2019**, *128*, 175–185. [CrossRef]
75. Schnabel, U.; Niquet, R.; Krohmann, U.; Winter, J.; Schlüter, O.; Weltmann, K.D.; Ehlbeck, J. Decontamination of microbiologically contaminated specimen by direct and indirect plasma treatment. *Plasma Process. Polym.* **2012**, *9*, 569–575. [CrossRef]
76. Tong, J.; He, R.; Zhang, X.; Zhan, R.; Chen, W.; Yang, S. Effects of atmospheric pressure air plasma pretreatment on the seed germination and early growth of andrographis paniculata. *Plasma Sci. Technol.* **2014**, *16*, 260–266. [CrossRef]
77. Fadhlalmawla, S.A.; Mohamed, A.A.H.; Almarashi, J.Q.M.; Boutraa, T. The impact of cold atmospheric pressure plasma jet on seed germination and seedlings growth of fenugreek (Trigonella foenum-graecum). *Plasma Sci. Technol.* **2019**, *21*, 105503. [CrossRef]
78. Alves, C., Jr.; de Oliveira Vitoriano, J.; da Silva, D.L.; de Lima Farias, M.; de Lima Dantas, N.B. Water uptake mechanism and germination of Erythrina velutina seeds treated with atmospheric plasma. *Sci. Rep.* **2016**, *6*, 33722. [CrossRef]
79. Silva, D.L.; Farias, M.D.; Vitoriano, J.D.; Alves, C., Jr.; Torres, S.B. Use of Atmospheric Plasma in Germination of *Hybanthus calceolaria* (L.) Schulze-Menz Seeds. *Rev. Caatinga* **2018**, *31*, 632–639. [CrossRef]
80. Molina, R.; López-Santos, C.; Gómez-Ramírez, A.; Vílchez, A.; Espinós, J.P.; González-Elipe, A.R. Influence of irrigation conditions in the germination of plasma treated Nasturtium seeds. *Sci. Rep.* **2018**, *8*, 16442. [CrossRef]
81. Pawlat, J.; Starek, A.; Sujak, A.; Kwiatkowski, M.; Terebun, P.; Budzeń, M. Effects of atmospheric pressure plasma generated in GlidArc reactor on *Lavatera thuringiaca* L. seeds' germination. *Plasma Process. Polym.* **2018**, *15*, 1700064. [CrossRef]
82. Štěpánová, V.; Slavíček, P.; Kelar, J.; Prášil, J.; Smékal, M.; Stupavská, M.; Jurmanová, J.; Černák, M. Atmospheric pressure plasma treatment of agricultural seeds of cucumber (*Cucumis sativus* L.) and pepper (*Capsicum annuum* L.) with effect on reduction of diseases and germination improvement. *Plasma Process. Polym.* **2018**, *15*, 1700076. [CrossRef]
83. Ji, S.H.; Choi, K.H.; Pengkit, A.; Im, J.S.; Kim, J.S.; Kim, Y.H.; Park, Y.; Hong, E.J.; Jung, S.K.; Choi, E.H.; et al. Effects of high voltage nanosecond pulsed plasma and micro DBD plasma on seed germination, growth development and physiological activities in spinach. *Arch. Biochem. Biophys.* **2016**, *605*, 117–128. [CrossRef] [PubMed]
84. Song, J.S.; Lee, M.J.; Ra, J.E.; Lee, K.S.; Eom, S.; Ham, H.M.; Kim, H.Y.; Kim, S.B.; Lim, J. Growth and bioactive phytochemicals in barley (*Hordeum vulgare* L.) sprouts affected by atmospheric pressure plasma during seed germination. *J. Phys. D Appl. Phys.* **2020**, *53*, 314002. [CrossRef]
85. Sery, M.; Zahoranova, A.; Kerdik, A.; Sera, B. Seed Germination of Black Pine (Pinus nigra Arnold) after Diffuse Coplanar Surface Barrier Discharge Plasma Treatment. *IEEE Trans. Plasma Sci.* **2020**, *48*, 939–945. [CrossRef]
86. Ambrico, P.F.; Šimek, M.; Ambrico, M.; Morano, M.; Prukner, V.; Minafra, A.; Allegretta, I.; Porfido, C.; Senesi, G.S.; Terzano, R. On the air atmospheric pressure plasma treatment effect on the physiology, germination and seedlings of basil seeds. *J. Phys. D Appl. Phys.* **2019**, *53*, 104001. [CrossRef]

87. Sajib, S.A.; Billah, M.; Mahmud, S.; Miah, M.; Hossain, F.; Omar, F.B.; Roy, N.C.; Hoque, K.M.F.; Talukder, M.R.; Kabir, A.H.; et al. Plasma activated water: The next generation eco-friendly stimulant for enhancing plant seed germination, vigor and increased enzyme activity, a study on black gram (*Vigna mungo* L.). *Plasma Chem. Plasma Process.* **2020**, *40*, 119–143. [CrossRef]
88. Sarinont, T.; Katayama, R.; Wada, Y.; Koga, K.; Shiratani, M. Plant Growth Enhancement of Seeds Immersed in Plasma Activated Water. *MRS Adv.* **2017**, *2*, 995–1000. [CrossRef]
89. Sivachandiran, L.; Khacef, A. Enhanced seed germination and plant growth by atmospheric pressure cold air plasma: Combined effect of seed and water treatment. *RSC Adv.* **2017**, *7*, 1822–1832. [CrossRef]
90. Lo Porto, C.; Ziuzina, D.; Los, A.; Boehm, D.; Palumbo, F.; Favia, P.; Tiwari, B.; Bourke, P.; Cullen, P.J. Plasma activated water and airborne ultrasound treatments for enhanced germination and growth of soybean. *Innov. Food Sci. Emerg. Technol.* **2018**, *49*, 13–19. [CrossRef]
91. Darmanin, M.; Kozak, D.; de Oliveira Mallia, J.; Blundell, R.; Gatt, R.; Valdramidis, V.P. Generation of plasma functionalized water: Antimicrobial assessment and impact on seed germination. *Food Control* **2020**, *113*, 107168. [CrossRef]
92. Naumova, I.K.; Maksimov, A.I.; Khlyustova, A.V. Stimulation of the germinability of seeds and germ growth under treatment with plasma-activated water. *Surf. Eng. Appl. Electrochem.* **2011**, *47*, 263–265. [CrossRef]
93. Takaki, K.; Takahata, J.; Watanabe, S.; Satta, N.; Yamada, O.; Fujio, T.; Sasaki, Y. Improvements in plant growth rate using underwater discharge. *J. Phys. Conf. Ser.* **2013**, *418*, 012140. [CrossRef]
94. Zhang, S.; Rousseau, A.; Dufour, T. Promoting lentil germination and stem growth by plasma activated tap water, demineralized water and liquid fertilizer. *RSC Adv.* **2017**, *7*, 31244–31251. [CrossRef]
95. Adhikari, B.; Adhikari, M.; Ghimire, B.; Park, G.; Choi, E.H. Cold Atmospheric Plasma-Activated Water Irrigation Induces Defense Hormone and Gene expression in Tomato seedlings. *Sci. Rep.* **2019**, *9*, 16080. [CrossRef]
96. Islam, S.; Omar, F.B.; Sajib, S.A.; Roy, N.C.; Reza, A.; Hasan, M.; Talukder, M.R.; Kabir, A.H. Effects of LPDBD Plasma and Plasma Activated Water on Germination and Growth in Rapeseed (*Brassica napus*). *Gesunde Pflanz.* **2019**, *71*, 175–185. [CrossRef]
97. Sarinont, T.; Amano, T.; Koga, K.; Shiratani, M.; Hayashi, N. Multigeneration Effects of Plasma Irradiation to Seeds of Arabidopsis Thaliana and Zinnia on Their Growth. *MRS Online Proc. Libr. Arch.* **2015**, *1723*, mrsf14-1723-g03-04. [CrossRef]
98. Hayashi, N.; Ono, R.; Shiratani, M.; Yonesu, A. Antioxidative activity and growth regulation of Brassicaceae induced by oxygen radical irradiation. *Jpn. J. Appl. Phys.* **2015**, *54*, 06GD01. [CrossRef]
99. Zahoranová, A.; Hoppanová, L.; Šimončicová, J.; Tučeková, Z.; Medvecká, V.; Hudecová, D.; Kaliňáková, B.; Kováčik, D.; Černák, M. Effect of Cold Atmospheric Pressure Plasma on Maize Seeds: Enhancement of Seedlings Growth and Surface Microorganisms Inactivation. *Plasma Chem. Plasma Process.* **2018**, *38*, 969–988. [CrossRef]
100. Khamsen, N.; Onwimol, D.; Teerakawanich, N.; Dechanupaprittha, S.; Kanokbannakorn, W.; Hongesombut, K.; Srisonphan, S. Rice (*Oryza sativa* L.) Seed Sterilization and Germination Enhancement via Atmospheric Hybrid Nonthermal Discharge Plasma. *ACS Appl. Mater. Interfaces* **2016**, *8*, 19268–19275. [CrossRef]
101. Amnuaysin, N.; Korakotchakorn, H.; Chittapun, S.; Poolyarat, N. Seed germination and seedling growth of rice in response to atmospheric air dielectric-barrier discharge plasma. *Songklanakarin J. Sci. Technol.* **2018**, *40*, 819–823. [CrossRef]
102. Mitra, A.; Li, Y.F.; Klämpfl, T.G.; Shimizu, T.; Jeon, J.; Morfill, G.E.; Zimmermann, J.L. Inactivation of Surface-Borne Microorganisms and Increased Germination of Seed Specimen by Cold Atmospheric Plasma. *Food Bioprocess Technol.* **2014**, *7*, 645–653. [CrossRef]
103. Bruggeman, P.J.; Kushner, M.J.; Locke, B.R.; Gardeniers, J.G.E.; Graham, W.G.; Graves, D.B.; Hofman-Caris, R.C.H.M.; Maric, D.; Reid, J.P.; Ceriani, E.; et al. Plasma-liquid interactions: A review and roadmap. *Plasma Sources Sci. Technol.* **2016**, *25*, 053002. [CrossRef]
104. Bourke, P.; Ziuzina, D.; Boehm, D.; Cullen, P.J.; Keener, K. The Potential of Cold Plasma for Safe and Sustainable Food Production. *Trends Biotechnol.* **2018**, *36*, 615–626. [CrossRef] [PubMed]
105. Schnabel, U.; Handorf, O.; Yarova, K.; Zessin, B.; Zechlin, S.; Sydow, D.; Zellmer, E.; Stachowiak, J.; Andrasch, M.; Below, H.; et al. Plasma-Treated Air and Water—Assessment of Synergistic Antimicrobial Effects for Sanitation of Food Processing Surfaces and Environment. *Foods* **2019**, *8*, 55. [CrossRef] [PubMed]

106. Yong, H.I.; Park, J.; Kim, H.-J.; Jung, S.; Park, S.; Lee, H.J.; Choe, W.; Jo, C. An innovative curing process with plasma-treated water for production of loin ham and for its quality and safety. *Plasma Process. Polym.* **2018**, *15*, 1700050. [CrossRef]
107. WIPO—Search International and National Patent Collections. Available online: https://patentscope2.wipo.int/search/en/search.jsf (accessed on 25 June 2020).
108. JP2019/030643 Rice Plant Production Method. Available online: http://www.freepatentsonline.com/WO2020027342A1.html (accessed on 25 June 2020).
109. JP2016009066A Animal and Plant Growth Promotion Methods. Available online: http://www.freepatentsonline.com/JP2016152796A.html (accessed on 25 June 2020).
110. RU02317668 Method for Treatment of Plant Seeds and Apparatus for Performing the Same. Available online: https://patentscope2.wipo.int/search/en/detail.jsf?docId=RU29606471&_cid=JP1-KBU5QY-51437-1 (accessed on 25 June 2020).
111. CN108738474 Seed Treatment Method of Selenium-Rich Black Beans. Available online: https://patentscope2.wipo.int/search/en/detail.jsf?docId=CN233991167&_cid=JP1-KBU5TM-53840-1 (accessed on 25 June 2020).
112. CN109041641 Seed Treatment Method for Improving Zinc and Selenium Content of Pumpkin. Available online: https://patentscope2.wipo.int/search/en/detail.jsf?docId=CN235612147&_cid=JP1-KBU5UW-55097-1 (accessed on 25 June 2020).
113. CN106817954 Method for Culturing Seedling of Oily Peony Seeds. Available online: https://patentscope2.wipo.int/search/en/detail.jsf?docId=CN199374287&_cid=JP1-KBU5W2-56424-1 (accessed on 25 June 2020).
114. CN103999593 Method for Breeding Wheat by Cold Plasma Treatment. Available online: https://patentscope2.wipo.int/search/en/detail.jsf?docId=CN107363409&_cid=JP1-KBU5WY-57195-1 (accessed on 25 June 2020).
115. CN107960254 Seedling Growing Method of Dalbergia Odorifera. Available online: https://patentscope2.wipo.int/search/en/detail.jsf?docId=CN215523929&_cid=JP1-KBU5XU-57795-1 (accessed on 25 June 2020).
116. CN104770103 Seedling Raising Method for Oil-Used Peony Seeds. Available online: https://patentscope2.wipo.int/search/en/detail.jsf?docId=CN151324684&_cid=JP1-KBU5YQ-58665-1 (accessed on 25 June 2020).
117. CN108243662 Plasma Preparation Used for Increasing Corn Yield, and Preparation Method and Application Method Thereof. Available online: https://patentscope2.wipo.int/search/en/detail.jsf?docId=CN223835065&_cid=JP1-KBU5ZP-59644-1 (accessed on 25 June 2020).
118. Sarinont, T.; Amano, T.; Koga, K.; Shiratani, M.; Hayashi, N. Effects of Atmospheric Air Plasma Irradiation to Seeds of Radish Sprouts on Chlorophyll and Carotenoids Concentrations in their Leaves. *MRS Proc.* **2015**, *1723*, mrsf14-1723-g02-04. [CrossRef]
119. Degutytė-Fomins, L.; Paužaitė, G.; Žūkienė, R.; Mildažienė, V.; Koga, K.; Shiratani, M. Relationship between cold plasma treatment-induced changes in radish seed germination and phytohormone balance. *Jpn. J. Appl. Phys.* **2020**, *59*, SH1001. [CrossRef]
120. Kyzek, S.; Holubová, Ľ.; Medvecká, V.; Tomeková, J.; Gálová, E.; Zahoranová, A. Cold Atmospheric Pressure Plasma Can Induce Adaptive Response in Pea Seeds. *Plasma Chem. Plasma Process.* **2019**, *39*, 475–486. [CrossRef]
121. Wang, X.-Q.; Zhou, R.-W.; De Groot, G.; Bazaka, K.; Murphy, A.B.; Ostrikov, K. Spectral characteristics of cotton seeds treated by a dielectric barrier discharge plasma. *Sci. Rep.* **2017**, *7*, 5601. [CrossRef]
122. Koga, K.; Attri, P.; Kamataki, K.; Itagaki, N.; Shiratani, M.; Mildaziene, V. Impact of radish sprouts seeds coat color on the electron paramagnetic resonance signals after plasma treatment. *Jpn. J. Appl. Phys.* **2020**, *59*, SHHF01. [CrossRef]
123. Ahn, C.; Gill, J.; Ruzic, D.N. Growth of Plasma-Treated Corn Seeds under Realistic Conditions. *Sci. Rep.* **2019**, *9*, 4355. [CrossRef] [PubMed]

MDPI
St. Alban-Anlage 66
4052 Basel
Switzerland
Tel. +41 61 683 77 34
Fax +41 61 302 89 18
www.mdpi.com

Processes Editorial Office
E-mail: processes@mdpi.com
www.mdpi.com/journal/processes

www.ingramcontent.com/pod-product-compliance
Lightning Source LLC
LaVergne TN
LVHW070631170726
843515LV00004B/226
* 9 7 8 3 0 3 6 5 4 3 1 9 2 *